For the Sake of History

The True Story of the 2020 Creek Fire

Jeff Young

Cover photo taken from Big Creek Bridge at 12:52 a.m., September 5th, 2020, looking across the Big Creek powerhouse #1 afterbay. The reflected light is midspan of the afterbay dam. Four hours before this, the fire was less than one acre. Credit: Nathan Crane.

design services provided by Linden Publishing

Printed in the United States of America.

ISBN: 979-8-9968736-0-9

Contents

Introduction

So much goes into a book of this nature; thoughts, words and photos are a small part of its construction. The contributors that must be thanked are not always obvious, not always acknowledged sufficiently, if at all, within the text.

Memories of people, places, and occurrences, both good and bad, become emotional focuses that still linger well after being put to print. The individuals from my past that deserve thanking for their spiritual contribution to this book are my grandmother, Ruth Young, whose strength and discipline were only overshadowed by her love and dedication to her family. My father, Stephen Douglas Young, whose seventeen short years in my life have given direction to my last fifty. These two embodied what is important in life and will never be forgotten.

My thanks also go to those that helped directly in making this book possible: my wife, Dina, whose English major was put to test in unscrambling my thoughts put to paper. Also, for her patience and tolerance for my absence during my time writing, and for the previous four years of frustration trying to make sense out of the first four hours of the Creek Fire. My 8th grade teacher, Chris Hardy, whose expertise and 50+ years of friendship made him the right person to critique my work. John Mount, who hired me at 14 and is still a close friend today whose wealth of information and knowledge I have always leaned on.

And last, the mountain community, which goes far beyond the boundaries of Fresno and Madera Counties. Without their contribution of information and support, I would have never gotten to this point. Stubbornness and gut feelings only go so far; it took the tidbits and bombshells that this community contributed to get to where we are now. And where is that? To the point where an independent, legitimate reinvestigation needs to be done - or a lawsuit is pursued against individuals or agencies involved. The mountain community deserves this.

Oh, what a tangled web we weave, when first we practice to deceive.—Sir Walter Scott, 1808

Ruth (Hauert) Young. A pioneer woman whose nature is still a part of our family. Credit: Young Family photo

Chapter 1

Let's Step Back 110 Years Before the Creek Fire

The Great Fire of 1910 killed 87 firefighters while consuming 3 million acres. Over the course of two days in mid-August of 1910, two thousand or more fires were burning in the inter-mountain area of the Pacific Northwest. The area had experienced two years of drought and persistently high summer temperatures, which allowed any fire ignited, whether by man or nature, to burn nearly at will. On August 20th, hurricane-force winds began to blow in advance of a major storm front coming ashore from the Pacific Ocean.

Many of the innumerable fires were being actively fought to protect towns and railroad structures, while others were being left to burn until the rains came and put them out. The nature of the winds over the next two days took nearly all of these fires - one estimate states nearly 3,000 individual blazes - and made them one. The burning together of so many fires over such a large area made escape extremely difficult. Towns, logging camps, and homesteads burned; people walked, rode horseback, or jumped trains to leave, only to find their routes blocked. Rivers, meadows, and rocky areas became refuges from the intense heat. A train overloaded with fleeing families hid in a tunnel to escape the flames. A group of firefighters took refuge in a mine shaft, the conflagration outside pulling air from within the tunnel leaving the trapped men with inadequate oxygen and struggling to breathe.

The rain from the approaching storm came two days too late to save the lives of many firefighters. This fire, also known as the Big Burn, is still the single most deadly wildfire in loss of life to firefighters in U.S. history. It is also the largest; its 3-million-acre footprint covered portions of Montana, Idaho, Washington, and southern British Columbia.

This fire brought awareness to the fledgling Forest Service of the need for federally based firefighting policies which were quickly crafted and funded. The 'extinguish all fires policy' was implemented at this time with the goal of putting out fires by 10:00 AM the next morning: a worthy goal but an unattainable reality. This aggressive challenge began the debate of safety of firefighters over the importance of homes and forests. In an attempt to fight fires with little more than shovels and buckets of water, even with sufficient manpower, firefighters' lives would be lost.

Fifteen firefighters died on the Shoshone National Forest's Blackwater Fire on August 18th, 1937, when trapped by a spot fire ignited by embers landing behind them. They had no means of escape and policies and protocol were not yet formulated to address this situation. Without training, protective gear, or tactics to facilitate survival, these firefighters could only run, hide, and attempt to survive.

The Mann Gulch Fire of August 4th, 1949, near Helena, Montana, took twelve lives when a spot fire encircled the men.

The Rattlesnake Fire in the Mendocino National Forest of California, on July 9th, 1953, ended fifteen firefighters' lives. Like those on the Blackwater and the Mann Gulch Fires, they were trapped by a spot fire behind them and had not identified or constructed a means of escape.

November 1st, 1966, a fire in the Angeles National Forest, California, takes another dozen lives. Again, wind-cast embers ignited a fire beyond the main blaze 700 yards below the firefighters and into the bottom of a 'chimney canyon.' By this time new safety strategies had been introduced into wildfire control procedures, two of which were safety zone identification and escape routes. These new protocols were instituted on this fire and failed the test as the fire covered the distance of seven football fields in less than a minute, cutting the firefighters off from their designated safety zone. The Loop Fire, as this blaze was called, brought about further changes in firefighting with new guidelines for fireline construction strategies, safety equipment, and fire behavior training.

Fourteen firefighters lost their lives on Storm King Mountain, Colorado, on July 6, 1994, by the same reoccurring event - entrapment by spot fires.

Nearly twenty years later another disaster struck the wildfire-fighting community. On June 30th, 2013, nineteen members of the Granite Mountain Hotshots lost their lives while fighting the Yarnell Hill Fire near Prescott, Arizona. They again were trapped by a spot fire without a means of escape. And once again, protocol was inadequate or unfollowed, proving fatal.

174 firefighter deaths on seven fires over 103 years. Firefighting, even with strict safety protocols, creates opportunity for disaster. When Mother Nature meets human nature in extreme fire conditions, this outcome can be expected.

If the reasons for these deaths were not investigated and documented truthfully and acted on properly, what could ever be gained as usable knowledge? More importantly, what could be lost? If causes of the fires were not established by diligent and honest investigations, how could culpability and accountability ever be established? Wrong information going into the historical record, whether fraudulent or not, could allow for the same results.

So, what is history without truth? A lie. Useless information at best, a dangerous corruption of history's lessons at worst. The same actions will produce the same results.

How does this relate to the 2020 Creek Fire, which burned 380,000 acres and 652 homes? Wrong information was placed into an official document and disseminated through the media, thus 'changing' history. What came from those in a position of trust was intentionally manipulated to protect themselves by deception. Little information from the first four hours of the fire can be trusted. The Creek Fire history, as we know it now, is a lie, a fraud, a conspiracy, a coverup. Fortunately, no one died because of this.

Several involved in the fire during this time were told not to talk and most have not, while some have talked and have been untruthful.

Also, CAL FIRE was denied access to the fire by the U.S. Forest Service who said they had authority over the fire. This decision produced devastating results. Many questions - no answers.

Not to be outdone by the questionable actions of the USFS, evidence points toward the Fresno County Sheriff's Department being on scene long before anyone else and long before the fire was reported by a resident. What were they doing there while not summoning a fire response? Did they have knowledge of the marijuana grow near the point of origin or the cause of the fire? Were they assisting someone because of this knowledge?

Why did those trusted to serve fail their mission? Were mistakes made? Yes. Were people on scene unorganized and unprepared? Yes. Were unqualified people in positions they could not handle? Yes. Were people there with prior knowledge of the marijuana grow? Likely. These reasons, though unproductive and potentially dangerous, unfortunately, must be expected and accepted; that is how life works. What is unacceptable is to willingly alter the truth to hide accountability. Breaking protocol is not uncommon and is often necessary and productive; to lie about what was done and not done in breaking protocol is wrong, and to then enter that fraudulent information into an official document is criminal.

Likewise, what's wrong with growing a little weed on government land? Considering the fact that this is illegal - which makes it covert to avoid discovery, and which produces hidden operations in the worst possible places for fires to ignite, while poisoning the environment - makes it terribly wrong. When a fire does erupt because of this, and then to have officials ignore material evidence that suggests findings contrary to those found in their official document, is beyond unacceptable. That, too, is criminal.

And then to avoid transparency for years when hundreds of landowners lost homes, hundreds of thousands of animals lost lives, and millions of trees lost value, whether commercial or spiritual; is that not criminal as well? This is what happened with the Creek Fire. The USFS Report of Investigation was signed as complete 7/15/21. It is a fraudulent document intended to divert blame from a human-caused fire. Yes, there was a marijuana grow with recent human activity nearby. No, there was not a lightning strike.

All sides of 'what a forest is for' should be enraged over this, from the, 'log it to save it capitalists' on the right to the, 'let it be and let it burn' socialists on the left. Transparency should be demanded, accountability identified, and justice served.

Chapter 2

The Fire Triangle

Speaking on something far from rocket science, I would like to explain the simple science of fire as it pertains to wildfires and campfires. All that needs to be known to understand and be proficient with fire is what humans have known for close to a million years. Anyone who seeks to make fire complicated is likely just uneducated in the simple equation that makes it work, or those in the business of wildland firefighting. This is not directed at those with boots on the ground. It is directed at those that make the exorbitant, unjustified salaries with nothing to show for their cost to the taxpayers except destroyed property and forest land. From regional fire management up to mayors, the governor, and even up to the president - many times they try to convince us of the complexity of fire while ignoring their mistakes, unpreparedness, underfunding, or waste. They are covering for their failures. They are blowing smoke. The fire triangle…no one understood it better than a caveman.

As the term suggests, just three things are necessary to equal fire - fuel, heat, and oxygen - and, as the fire triangle itself suggests, these components must all be present. Take any one away and you won't have fire. It's that simple. It's that easy to understand.

For a caveman, in his routine to survive, fire was more than a convenience. Food was best if cooked and often required cooking, snow and ice could quickly be converted into an adequate supply of water, and shelter was made much more tolerable/survivable with a warming fire.

In any fire, fuel is first concern, and the drier the better. Of utmost importance to the fire-builder was the initial fuel, called tinder. This can be shown no better than with Ötzi, the Copper Age hunter (5,000 years before present) whose body was found in a European glacier. One of the three things carried in his pack relating to fire-building were pieces of fungus tinder, his initial fuel. He realized the importance of this component and tried never to be without it.

In the wildland fire setting, just like Ötzi's campfire, the drier the fuel the easier it ignites, the hotter it burns, and the faster it moves through the available fuel. Fuel is the only component of the fire triangle that can be realistically controlled to an effective degree at the wildland fire scale. This is accomplished by fuels management; something that has only been limited by expense and controversy, and thus, we have Creek Fire devastation.

The Southern California fire tragedies, in urban/metropolitan/forest interfaces, were no different. How much money was spent on the unimportant while underfunding fire prevention and response? In the end, how many endangered creatures saved by regulated fuels management practices - which generally means doing nothing - are now dead from incineration or toxic runoff? Just like for Ötzi, fuel is a major concern for wildfire managers and must be addressed before a fire starts, which seems never to be done.

Ötzi was obviously very mobile as shown by the few necessities he carried. The other two items relating to fire found in his pack were there to provide the second leg of the fire triangle. He carried a birch bark container that held maple leaves, which was the insulator for the glowing coal that would be transported - sometimes for many hours - to his next fire. There it would be gently coaxed to flame once more.

In the wildland fire setting, there are only six possible natural means to produce the heat necessary to ignite the fire triangle. These are: meteor strikes, sparks from rockfalls, coal seam fires, spontaneous combustion from composting organic material, volcanic eruptions, and lightning. Of these, I have never witnessed meteor or rockfall fires, even on YouTube. I have seen burning coal seams and volcano-caused fires on the news. In contrast, I have witnessed the two other natural fire causes numerous times in my day-to-day life. Spontaneous combustion is still frequently the cause of fires in large sawdust piles and compost operations. Lightning is a frequent cause of fire in the western United States.

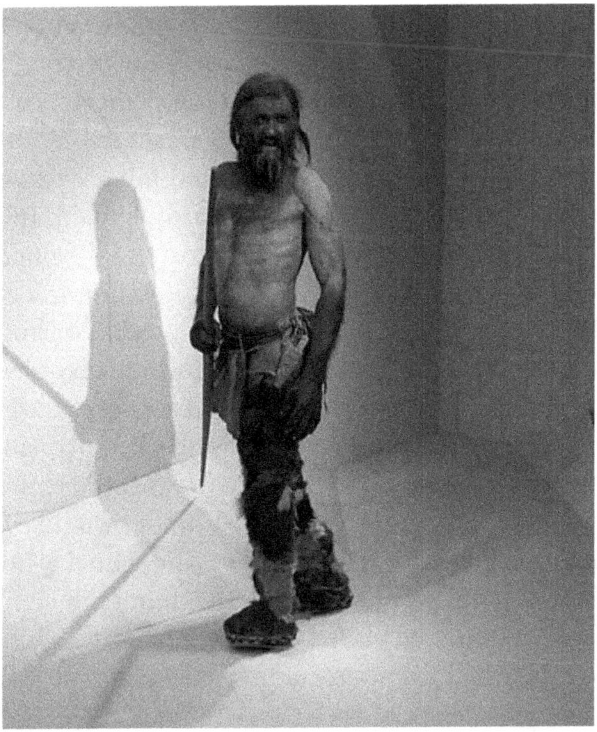

Reconstruction of Ötzi located in the South Tyrol Museum of Archaeology, Bolzano, Italy. Found Melting out of an Alps glacier in 1991 where he had been entombed for 5,300 years. His cause of death was from a flint-tipped arrow whose point was found lodged within his shoulder. Credit: Wiki Commons

As to the wildfire and the campfire, we have covered fuel and heat. To complete the equation oxygen must be introduced to the fire triangle. Oxygen is the lesser component of the two primary elements comprising the air we breathe, with nitrogen being the other. Oxygen, more than sufficient in the air around us to do its job, is required for a fire to ignite, survive, and spread. Ötzi carried this third leg of the triangle, not in his survival bag, but in his lungs. After gathering fuels from his current environment, from moss and grass to twigs, limbs and logs, he would begin the necessary survival practice of building a fire. A dry base was important; if the ground was damp in any way, a base of dry bark, rock, sand, or anything available that had been protected from the elements, could be used. It was not necessary that it be flammable, just dry. Atop this would be placed a layer of grass, moss, shredded inner bark, or anything available that was light, thin, dry and easily ignited. Into this light fuel, carefully unwrapped after being transported from his last fire, would be placed his valuable ember. His attention to detail had assured him that this ignition source was protected during his travels in such a way that just enough air got to the barely living coal to allow it to 'survive,' but not so much that it consumed itself within its insulating wrap. This package was likely checked occasionally to see that it remained viable. Placed in contact with the ember and then blown on, his faithfully protected fungus tinder would eventually ignite from the heat of the coal and Ötzi's 'wind.' Quickly but delicately, the smallest of twigs would be fed to the minuscule fire until larger and larger fuels could be placed without stealing the slowly building heat. Too much fuel too soon could snuff the infant flame as easily as water.

In the wildland fire setting, wind is the driving force. Nothing more needs to be said to anyone who witnessed the Southern California fires or the Lahaina Fire. Without wind, a quick, well-orchestrated response supported by sufficient resources is generally sufficient in controlling fires in a reasonable amount of time, with an acceptable amount of loss. Add wind of any strength and firefighting becomes much more difficult. Winds above 40 mph make fire the advancing army that must be defeated with force, thought, management, and resources - or run from and left to pillage. Southern California and Lahaina were lacking in all of these aspects of the job and lost the battle, not by lack of knowledge, but by ignoring standard firefighting requirements, such as assuring the fire is completely out before leaving. Both the Palisades and Lahaina fires were thought to have been extinguished before rekindling by the 'breath' of a strong, dry wind.

In other words, it is not the fault of the boots-on-the-ground firefighters that Southern California or Lahaina burned, but that of the bureaucracy that cares only about itself. The firefighters were abandoned before the fight and thus forced to run. That is the story of these fires - and of the 2020 Creek Fire.

With this sufficient understanding of fire and its bureaucracy, the story I will now tell, that of the Creek Fire, may be more meaningful. I shall begin.

6:33 PM, September 4, 2020, a forest fire is reported in the mountains of Fresno County, California: The Creek Fire. Who could have predicted that this lazy little smoke would turn into the largest wildfire in California history? People that witnessed the 1994 Big Creek Fire might have. I was one of those.

The Big Creek Fire, a short-lived burn of 5,000 acres, started when a squirrel exploring a down-canyon powerhouse substation crossed two exposed electrical conductors. Whether it was thrown by the shock arcing through its body, or ran burning into the grass beyond the fence, the flaming rodent ignited dry grass that would soon become the Big Creek Fire on the afternoon of August 24th, 1994.

Once this rodent-caused flame reached the dry grass, the fire took quickly, then steadily made its way up canyon in the thick brush beyond the grassy slope the substation had been carved into. Late day heat and prodding breezes, with the occasional gust, was all it took to move the fire out of reach of any fire response other than that of aircraft. This airborne response arrived within the hour and did its job in slowing the advancing flames, but the coming darkness would halt this attack and ground crews were not yet able to back up the aircraft. Road access to the far side of the creek where the fire burned did not exist, and the terrain was too dangerous for hand crews to navigate after nightfall. Further engagement would wait until morning. The fire would be left to the whims of nature as evening approached and would move about as fuel and winds saw fit.

The stronger afternoon winds would first dwindle to light thermals as the sun settled in the west that evening and should eventually change to a down-canyon breeze with the coming of darkness - the normal wind routine for the area. The old-timers watching the flames from several miles upstream knew this and said the eventual change would limit the fire's progress and allow it to be contained first thing in the morning, 'before the winds woke up.'

Even with the optimism of the old-timers that the fire would be of no threat, I still decided I would travel the short distance from our vantage point to the east edge of town with my chainsaw, axe and wedges, and begin falling dead trees. As I was a timber faller by trade, this bit of effort was second nature to me. It was just a precautionary measure that I felt was prudent - removing a stand of dead, volatile trees that could pose a threat if the worst happened - and not a repudiation of my elders. I, too, believed the fire would be stopped by the next morning.

I began putting trees on the ground no more than a stone's throw below the most vulnerable portion of this small community - Jungletown as it was called - a quaint private section of the predominately power-company-owned town. Besides thinking it prudent, I had another reason to cut the dead trees. Being a timber faller and seemingly born to be one, I needed little excuse to cut a tree down. I had not cut timber that season as logging was stagnant, limited by the spotted owl regulations that were becoming stifling. Instead of cutting timber for a logging company by traveling who knows where to find work, my current means of employment was teaching science at the nearby elementary school. With work in the woods, as logging was called, so scarce, I would do this for the school year and see what next logging season would bring.

By darkness I had put nearly three dozen Ponderosa pines on the ground - a patch of trees killed by the ever-present bark beetles - and returned to join most of the town, which had migrated to 'The Point' to watch the flames. The Point, named decades ago, is the solid rock end of the nearly solid rock ridge that Big Creek was built upon over 80 years before. There is an incredible view from here even when there is nothing exciting to look at; the perfect spot to watch a distant lightning storm, falling stars, or, as the gathering group of locals were doing this evening, watch an approaching forest fire.

The audience was there because it was exciting and not necessarily concerning, most having the expectation of the fire's progress slowing to a creep. After all, that's what the old-timers were saying would happen. The fire would be pushed back upon itself by the changing wind and be a low level, lazy mass of smoke by morning. For most, watching the fire was more a social event than anything else, much like the 4th of July. Beers were consumed and stories were told, some about past fires, while kids played and teens flirted. Eventually the distant flames were little more than a backdrop to small-town gossip. No one was too concerned. The fire would be contained by lunchtime tomorrow was the consensus.

By ten o'clock that night there was apprehension in the air, along with much smoke. Youngsters had been ushered off to dinner and then to bed, out of earshot of the plans bantered about by adults, which were now fully focused on the flames. Those who had gone home for dinner, mostly mothers, returned for a last peek before sleep. They were not only surprised and disappointed, but frightened. The wind had not changed and the fire had not slowed; half the three-mile distance from its starting point to town was burned or burning. There were no aircraft, no bulldozers, no sweating firefighters' bodies between them and the flames, just more of what was already burning quite aggressively: dry brush and oak - and there was smoke, lots of smoke.

Through the night the fire stayed to the north of the creek the town was named for, but kept an unbroken, unhesitant front of marching flames slowed only slightly by the nighttime cooling. Big

Creek was not the only community it was advancing on. In the next few hours it would be known if the fire would cross the creek into Camp Sierra, an historic village of mostly summer cabins. Like Big Creek, Camp Sierra could only be protected at the edge of town. There were no roads between them and the fire - no line of defense. Soon fire trucks would be rolling into both communities. Few adults slept that night.

With first light, and the flames still approaching, plans made the night before were being instituted. Camp Sierra appeared safe for now, with the fire still across the creek and burning upstream, skirting the little resort and the several dozen cabins that comprised it. Air attack could now fly and hand crews could now safely deploy. At Big Creek, combustible debris were being removed all along the exposed homes to the north of Point Road. Folks from all over town came to help.

Our house sat on this north side about halfway between the elementary school where I worked and the firefighters' barracks at

The school-owned house my family lived in during the 1994 Big Creek Fire burning in the 2020 Creek Fire. 40 other homes were burned that day - nearly half of the homes in this small mountain community. Credit: Fresno Fire Department.

the heliport, with the landing pad being the last structure before the ridge dropped quickly away. As others raked and shoveled, I felled all the trees behind our rented, school-owned house. I made these fall directly away from the property and down the hill. This took the worst of the threat further from our home, as these green, resin-rich cedars would spread fire quickly if engulfed. They would have to remain lying on the ground unlimbed. I should have cut the limbs off to reduce their flaring if ignited but did not have the time. I had been requested to cut other trees and headed the half mile back to Jungletown to fall a massive pine tree that stood next to the school secretary's house. If fire caught hold of the needles in this tree, even though green, it would become a crowning inferno. Her and her husband's house would not be the only casualty.

Crown fires are the firefighter's worst dream, an unmanageable enemy, and that would be exactly what this tree would become: a 160-foot explosion of flame if ignited, spreading embers far into the neighboring homes. Jungletown could disappear in minutes. With the big pine on the ground, I was back home to reevaluate my own problems. I had decided that placing sprinklers behind the house would be a good idea, as it only took an ember to start a spot fire, and we had already watched several start in front of the main fire that day.

Upon returning to Point Road, I discovered the fire was beginning to threaten the westernmost homes. It was midday now and the up-canyon wind was building. Within minutes the fire blew up. Where it burned the previous three miles in the better part of a day, it consumed the next mile in less than an hour. This inferno ravaged the canyon wall opposite Big Creek to the north, skirting Jungletown and burning everything to the ridgetop in minutes. Its next run was toward Point Road.

By the gate to our yard hung red and white striped flagging, placed there by firefighters to designate that all houses beyond this were to be sacrificed once the flames entered the town. The fight could start, and possibly end, right there at the candy-cane ribbon. I quickly loaded my truck, inside and out, with valuables and sentimentalities and proceeded to limb the fallen trees I had felled earlier. Then I started our only sprinkler.

By this time the street was lined with fire trucks, and a Chinook helicopter began dropping water at regular intervals, dipping its bucket in the nearby hydroelectric powerhouse discharge bay. Back and forth it went taking less than two minutes to a loop, dropping nearly three thousand gallons of water directly on the burning brush and trees each time. Even with this efficient firefighting machine's attack, the fire was winning.

Big Creek lies at the transition zone between brush and timber. While overnight the fire burned through mostly brush, which back-lit an occasional bull pine as its resinous needles exploded into flags of energetic flame and oily black smoke, it now burned in mostly Ponderosa pine trees with a heavy brush understory - an explosive combination often resulting in crowning fires. This was the fire the helicopter crew now fought. Within 30 seconds of each drop, the fire had regained its strength and raged on, climbing into the crowns of tree after tree from the brush below and spreading from crown to crown where the trees were close enough to ignite their neighbors. This exponential spread was only limited by each returning bucket of water that would target the most threatening mass of flame. It appeared the candy-cane ribbon would become the point of attack, and all homes beyond would be allowed to burn.

Between two of these repetitive water drops, an airtanker painted The Point from 200 feet above with slimy, neon red retardant, covering homes, yards, Point Road, and vehicles - mine included. (In 2020 I finally parted with a favorite down jacket placed in the back of my truck that day, 25 years earlier. It still carried the faded pink stain of fire retardant.) For about 30 minutes things looked hopeless, but the fire was kept at bay by the helicopter and only fingers of flames burned into the backyards.

Our house, and likely a dozen others, was saved that day by that big black airship. It was only chance that brought it to Big Creek that afternoon. It had been purchased from a wealthy person on the east coast just days before by Columbia Helicopters, a logging and firefighting company based out of Aurora, Oregon. It was in transit there when it was redirected to Fresno, fitted with a line and fire bucket, and sent to the fire. It arrived as the fire crowned and its relentless airdrops saved the houses on Point Road. The previous owner of the airship? Donald Trump.

By nightfall the Big Creek Canyon was a mass of smoke, the wind was nearly still, and the fire front had moved beyond the canyon wall and out of sight toward Huntington Lake. All around the

My father, Stephen Douglas Young, logging in the Shaver Lake area the summer before he enlisted in 1942. Credit: Young Family photo

north edge of Big Creek, when the smoke permitted, flaming stumps and snags could be seen. The local power company, Southern California Edison Company, had used its small bulldozer to put in a fireline below Jungletown, which steered the flames just to the edge of the homes there, where garden hoses, firetrucks, and hose lays from distant hydrants had knocked the flames down. The big pine tree I had cut lay partially burned, whether cutting it made a difference, who knows? What was important: not a home was lost, nor a person injured.

Twenty-six years later, a different story would begin.

Chapter 3

Fire Memories

I grew up in a fire-conscious social environment; everyone that lived in the mountains did to some degree. Most thought about fire concerns only occasionally, when something was already burning or a thunderstorm was rolling in. They would then prepare and react, if necessary, or give the thought some lip service and go about their business. It was on the mind of a few from the beginning of every fire season to its end, and these citizens were always ready in both individual preparedness and for a community fire response. My dad was one of those.

My first fire memory is one of the few old memories I still retain, hanging in the closet of my mind to be found when needed, but otherwise forgotten for years on end.

The fire was just a small smoke when I first became aware of it, spotted by my father out of the corner of his eye. He was good at that. It was in a place called Jose Basin, a brush-choked bomb sitting below our small mountain community of Shaver Lake. The Basin, as it was referred to by the locals, was just waiting for the fuse to be lit. It had not seen a good burn in anyone's memory.

I was just old enough to understand the simplest facts of fire then, thanks to our ever-hungry fireplace constantly consuming firewood. That fire kept us warm when fed sufficient wood and occasionally burned the careless. It was as popular as the Wonderful World of Disney in the winter months but hidden until fall by the various summer needs placed upon its hearth come spring. I would be introduced that evening to a greater fear of fire than a fireplace casting sparks; that bit of smoke in the distance portended potential disaster.

We had gone to the house of some close friends a few miles down the road toward Pine Ridge to eat dinner and visit. Their small summer cabin sat in the midst of an eighty-year-old apple orchard from which the ground dropped steeply to The Basin, 2,000 feet below. About the time dinner was served was when my father saw the wisp of smoke. He had developed this knack over years of chasing fires in the mountains. He owned a 250-gallon fire pumper setup on the back of a Willys pickup truck which was contracted locally by the U.S. Forest Service for firefighting. For years he was often the first on scene to mountain fires. His goal was to beat the U.S. Forest Service -'The Feds' he called them - to the fire.

The smoke below, to me, was insignificant, but to my father it was of great concern. He spoke of this being the one; the fire that could race 'up the chimney' to Shaver. This talk frightened me, but not as much as his body language that even a child could understand. We focused like hawks while the smoke grew and flames became visible. With darkness we could see the red flashing lights of fire vehicles converging on the flames of the fire. One of the vehicles had gotten there much more quickly than the others, likely because someone was in the right place at the right time. My father pointed out to those listening that a quick response is the greatest prerequisite to success in putting out a fire. Get there quickly and an individual might put it out; get there late and an army might not.

For a short period of time the talk was of evacuation and where we would go and what we would do. My father would head for the firehouse and start the siren to notify the other volunteers of the threat, but only if the fire began a run at town. If this happened, my mom and I would locate my siblings and we would all go to my grandparents' house and wait for what the fire might bring. She was directed to make sure the car - a rather worn '56 Chevy station wagon - had gas, and to get sleeping bags and food into it. We would flee the fire toward the nearby lake if the fire began to enter town. Going down the hill toward the foothills could get us trapped by the flames; going up the hill gave us miles of escape options.

Fortunately, it became evident that this would not be necessary. Although the brush was tinder dry and impenetrably thick, a network of roads in The Basin served as both access and firebreak. The

Fire on the backside of Musick Mountain as seen from near Alder Springs the evening of September 5th. Jose Basin is in the foreground and will be consumed by flame over the next two days while the fire threatened Shaver Lake and consuming several hundred homes in Alder Springs and the Pine Ridge area. Credit: Karen Tillison

fire was contained relatively quickly that evening. Two things made this possible: a quick response and the missing ingredient - wind. I stayed up later that night than I ever had before. My father would not leave until flames were no longer visible. Besides, it was a good excuse for him and my mom to have another drink or two with close friends. That experience, over 60 years later, is still quite clear. Both the vision and the fear.

As I got older my knowledge and interest in fire grew, and I was always asking if I could go on a fire call. I was finally allowed to go when I was old enough to do as I was told and to do so quickly. My first fire response was to an escaped campfire near Dinkey Summit, about ten miles from town. The race was on. The response vehicle was now a '64 Jeep pickup, appropriately red in color. My dad was certain that we would beat 'The Feds' that evening.

We were quite close before the smoke could be seen. Being adjacent to a creek, the moist air and damp earth kept the duff, a ground-covering combination of shed deciduous leaves and evergreen needles, bark, and limbs, from burning rapidly. Years of visitors wandering around the camp and wildlife coming to drink from the creek had trampled this covering down to where mineral soil was exposed here and there. It was obvious this would not be much of a fire fight. The burning duff that we attacked could hardly be called a fire. The limbless trunks of trees around this camp could not carry the fire into their crowns, and brush was scarce. A few buckets worth of water out the end of the hose and the fire was nothing but smoke. By the time 'The Feds' arrived there was little left but steam. My dad turned the incident over to the uniformed 'official' firefighters, which meant getting in the truck and driving away with little more than an acknowledging hand gesture short of a wave. There were some attitude anomalies between my dad and 'The Feds' that precluded any comradeship. Was this because he won the race more often than not, or could it have been the steam he left them that did not smell quite right? My father also pumped septic tanks with his firefighting rig in the off-season and between fires. Whatever the reason, there was little love lost between him and 'The Feds.'

Chapter 4

A Bad Fire Year

A bad fire year was always the topic when summer came around and still is. Whether a statement by someone knowledgeable, or a question from someone not, the type of fire year was always discussed. It's a fact of life in fire country - people talk about it just like the weather. What makes a bad fire year? Too little precipitation or too much? Too hot? Too windy? Too dry? Yes, all of these things can contribute to a bunch of acres burned. What is the parameter for a bad year? A couple million acres burned, a couple thousand structures destroyed, or a dozen deaths? It varies, depending on who is reporting and their focus on value. Generally speaking, acres burned gets the most press and is the historical comparison; size seems to matter most after loss of life and property is forgotten. What is the one variable that always seems present in a bad fire year? Lightning. The years without this natural spark are rarely seen in the worst-years-ever list. You can have a truckload of fireworks, but without an ignition source it would be a boring 4th of July.

2020 was going to be a bad fire year. Then it wasn't. January and February that year were historically the driest ever in California. The future looked grim for a couple million acres of foothill and forest, then March and April came with above average precipitation and the predictions of many acres burning subsided. Into summer things remained normal with enough moisture to produce a good crop of grass to feed the cows, and a good growth of brush to feed the fires. Both cattlemen and firefighters would be happy. Maybe not overly happy, but happy. A profitable year for both seemed likely.

A bad fire year is good money for firefighters, a fact of the business that conflicts the minds of many in the fire industry, but a situation that is unavoidable. Four million acres burned means a lot of overtime with hazard pay. Should anyone feel bad about profiting from a fire disaster? Not if they do their job and contribute, even in the smallest way, to quickly extinguishing the fire. Most, but not all, see it this way. The boots on the ground will do what they are asked and get the job done as best they can, almost without exception. But this attitude of getting the job done seems lost upon management, who, more often than not, sees a fire put out tomorrow better than one put out today. This reality is often the result of the decision-making process, which is rarely quick in any government undertaking. In fire, time not fighting it is lost opportunity in extinguishing it, but for some, time not fighting it allows it to grow and become more profitable at hazard pay rates. These opposing values should not exist as a choice. The selfish side of human nature should never come to bear on the decision as to when a fire should be fought or how. I no longer look at the overall cost of a fire with much interest, but more toward who made money on the fire and how, and what aspects of the fire were neglected and why. This thought process identifies waste that could be better utilized, failures that could have been averted, and who deservingly, and undeservingly, made money. That is how I now look at a forest fire.

Fire is an industry, and the intent of those that profit is to grow the industry. This generally doesn't mean more planes to fight with, or less fuels for the fire to feed on. More often it means more bureaucracy. Acres, homes, and lives lost should be considered before growing the bureaucracy's size and control. Bureaucrats don't extinguish fires. More often than not they fuel them. Manpower, equipment, and sound practices put fires out. And a little forethought in managing fuels can keep big fires from ever happening.

July's fireworks often start the fire season. By this time the grass and brush is dry enough to ignite easily and ignition sources are in the hands of tens of thousands of patriotic revelers and a few fire-starting terrorists. Whether legal or not, in the tinder of early July, colorful sparks will do more than entertain. By the end of the month, large fires can be expected and lightning will replace fireworks as the primary cause. Lightning: Nothing new. Unavoidable. Unpredictable. It's been around for eons igniting the fires that recycle carbon. It is hard to forecast in general and even harder to predict as to when, where, and how much will strike the ground. But when it happens, there could be fires by the scores - all at once.

Prior to 1850 and the influx of gold-seekers there were fire years much worse than what we experience now in the terms of acres burned. A normal year would see 4 million acres burned - a 'bad' year would see 12 million. 12% of the land area of California would burn and no one gave a damn; the indigenous peoples of the forest would just move out of the way of a fire, which they may have intended to start anyway, and return to its footprint to hunt and gather in the luxurious regrowth of the years that followed. It was natives, intentionally or not, or lightning that caused fires back then. The only bad fire to an Indian was the one that went out before dinner was cooked. Fire was an important part of their existence. It was regenerative. It was part of the natural order. An order that can no longer exist naturally without devastating results.

Then came August, 2020. The weather pattern changed and brought what should have been a warm, moist disturbance to the coast of California. This subtropical air interacted with the hot dry air it had pushed against the Sierras in its advance. These lines of interface, where the two contrasting air masses met, caused extreme turbulence and towering clouds, which produced extensive lightning and the potential for fire-snuffing rain. But this year, the copious amounts of rain needed to limit fire starts did not materialize. Thousands of strikes arced between ground and clouds over a large area of the state, but the usual widespread downpours that accompanied these thunderstorms were nowhere to be found. Convection was so great that precipitation struggled to escape the clouds, and the air was so dry below the clouds that little rain reached the Earth's surface before evaporating. These conditions produced dry-lightning (down strikes without rain) and were nearly as likely to start a fire as not. At least it seemed that way to the responding firefighters. Thousands of dry-lightning strikes produced hundreds of fires during the thunderstorms of August 16-17, 2020.

Fausto, a tropical storm that existed for only six hours before degrading back to a tropical depression, was about to leave its mark on history. Located 1,000 miles off the tip of Baja California in the Pacific, 1,400 miles from the northern coast of California, and lasting less than a week, who would have ever thought this brief storm could be a distant purveyor of destruction? The moisture streaming north and east from the quickly dissipating storm came ashore in the San Francisco Bay Area early on the 16th and began interacting with the very hot and dry air over the land. Over the next two days, 11,000 lightning strikes would become the point of origin for 650 fires. The ease with which they burned is evidenced by the 38 individual fires that combined to form the August Complex, which burned parts of seven counties and three national forests; a million-acre behemoth that could fit the state of Rhode Island within its perimeter and the largest single fire scar in California history.

This rare atmospheric event lighted California up like never before, historically speaking, with nearly 2 million acres uncontained at one time. Within a week another disturbance was forecast to wreak havoc on the north state with abundant lightning, but, fortunately, rode the Easterly Jet Stream across the state without contributing much more than some erratic winds and an occasional downstrike; a great relief to the over-extended fire community and its limited resources. A week and a half with little lightning produced needed optimism for fire managers. Resources would gradually become available as containment grew on the many fires. This, along with several large fires burning into each other, thus hastening their containment, produced a feeling that things were looking up.

6:33 pm, September 4th, 2020, Big Creek, California

A plume of smoke is reported to U.S. Forest Service Patrolman 34 stationed at the barracks adjacent to the heliport at the end of Point Road in Big Creek. This begins the official response to what would become, briefly, the state's largest single ignition wildfire.

The first photo taken from Big Creek Point, 6:40 PM, September 4th, 2020. The fire remained small for the next 2 hours and 20 minutes before winds picked up. Two airtanker retardant drops and two helicopter water drops missed their target. This was acknowledged over the USFS radio recordings. Credit: Nathan Crane

Within minutes a full response has been requested and all available crews and equipment ordered. The potential for this small, harmless-looking column of smoke was not lost upon Patrolman 34; you could hear it in his voice over the radio. He was concerned, and he was alone as far as there being USFS firefighting personnel in the area. His small pumper truck, with the location of the fire, could not respond. Protocol required that safety guidelines be followed; driving down a marginal 4WD road into a brush-choked canyon would never be considered safe. That is one of many ways firefighters could die. Instead, he and the Forest Service Silviculturist, who lived in town as well, would be in charge of the response until higher rank relieved them. With no other options, all they could do was watch and relay information to the responding resources.

The normally occupied USFS barracks were empty of the Engine 32 crew members who were elsewhere in the heat of the far-ranging wildfire battle. Not that it would have made any difference. Even though being only 15 minutes from the fire, they too would have been at the dictates of protocol. With this safety protocol so important, why then did the first engine on scene, USFS Engine 42, which arrived 45 minutes after the first report of smoke, break this important safety precaution and drive the marginal 4WD road to within 300 feet of the fire? With a full tank of water and a full crew, the fire could have been easily contained, but no attempt at fighting the fire was made. Why then did the engine turn around and leave without engaging the fire? Did this happen because the fire captain was a trainee - the certified captain being on days off - and had realized his error and chose to leave the burning snag burning? Or was it because the evidence of a marijuana grow was found in the few minutes he was outside the engine 'putting eyes on the fire' at the request of the newly arrived Incident Commander, also a trainee? The lack of answers to these questions lies at the heart of the coverup of when, where, and how the Creek Fire was started and why it was not approached and put out quickly.

By ultimately blaming the fire on lightning, which was done, no further questions needed to be asked; no other answers needed to be sought; no one involved became accountable - only Mother Nature. It is stated by the silviculturist in the Report of Investigation (ROI) that this snag, which could have been extinguished by 7:30 PM, albeit after breaking protocol, was seen shedding the wind-blown embers that spread the fire one and a half hours later. This was not the case. A photo taken at about 7:50 PM shows no flaming snag. It had already burned out. What the photo did show was a small ground fire of approximately 500 square feet and two other light sources that could have been headlamps or flashlights.

I know one reason why E42 left the snag burning. The engine and crew should have never approached this fire due to safety rules and realized this after committing to the narrow 4WD road. They needed to get out of harm's way fast, so they left. The fire was burning in a steep canyon prone to high winds (two miles upstream is Huntington Lake, one of the best sailing destinations in the state). One-hundred-degree temperatures in the valley below that day could have quickly sent a hot wind up canyon to fan the flames, which would spread quickly through the desiccated fuels (moisture content was comparable to kiln-dried lumber). Another reason they left without engaging the fire - the disappearing pot grow?

Chapter 5

The Sheriff's Secret Fire

The official report of smoke, the one that produced a response of firefighters and equipment, was not the first report of the fire. Exactly 15 minutes earlier a 911 call was received by the Fresno County Sheriff's Department from a local resident reporting the fire. This is documented in the U.S. Forest Service's official report (ROI). What is not documented is why this call did not produce a fire response of men and equipment. There is evidence that CAL FIRE was notified at 6:18 PM and again at 6:22 PM of a fire near Camp Sierra by the Fresno County Sheriff's Department, but no response followed. Why did CAL FIRE not respond immediately and why was the USFS not notified at all? The fire was on national forest land and thus the Forest Service's responsibility. This is likely the reason for the lack of response by CAL FIRE and also the reason why CAL FIRE was not allowed to fight the fire when they did arrive.

Another question which needs to be answered by the Fresno Sheriff's Department is why in the minutes before the 911 call, two residents who attempted three 911 calls from two different cell phones received busy signals. These residents then called a neighbor who reached 911 by landline which produced the 6:18 PM 911 call to the Fresno Sheriff's Dispatch Center referenced in the ROI - which then did not relay this information to the USFS. It appears as if the

Fresno Sheriff's Department was hiding the fire's existence. Were there sheriff's employees in the area at the time of the fire? It is clear that a sheriff's employee that owns a cabin in the area was there, but was there any other sheriff's presence there? It now appears, discovered five and one-half years after the fire started, that Fresno County Sheriff Margaret Mims was also in Camp Sierra that first afternoon of the fire and was secreted away from the area. Why was it necessary to hide this information? Why could she not have any casual association to a pot grow and fire? Is there more to her story that necessitates absolute secrecy?

At this point in my exploration of information - I should not call it an investigation as I am not an investigator, a point made clear to me by more than one person in law enforcement - the water becomes murky. A radio call was made by a sheriff's employee at 6:22 PM stating a fire was near Camp Sierra. This should have initiated an immediate engine response under any conditions but did not produce one even under conditions of extreme fire hazard. Instead, it was up to someone else to see that the fire was responded to.

It was this radio traffic, heard by a volunteer fire chief over a scanner and relayed to a firefighter in Big Creek for confirmation, that produced the deployment of equipment. While waiting for verification of the fire, the fire chief also called the Fresno Sheriff's non-emergency phone line to ask why a dispatch had not yet been initiated. This call ends with the fire chief being hung up on by an unconcerned sheriff's employee whose best answer to the question was 'it will be looked into.' The fire was verified as a single column of lazy smoke and reported to Patrol 34 stationed at the USFS barracks. Time: 6:33 PM.

What were the sheriff's department employees doing there - if they were there, which evidence suggests they were - at about 5:45 PM? The evidence that places them there: Multiple people in various locations along State Highway 168 saw three vehicles pass by them, traveling together at a high rate of speed and heading east, which is in the direction of Huntington Lake Road and Camp Sierra. All descriptions were the same: white Tahoe-like vehicles with darkened windows and multiple antennas. A Facebook post even asked if anyone knew where the sheriffs were going in such a hurry.

Along with the question of who was in these vehicles, how did these unmarked vehicles get to Shaver Lake at about 5:30 PM? They would have had to have known of the reason for the high-speed response by no later than 4:45 PM in order to be seen in Shaver Lake when they were. How did the drivers of these three vehicles become aware of a need for them 40 miles away and what was that need? This was the Friday evening before the start of the three-day Labor Day weekend. If sheriff's vehicles were seen going somewhere in a high-speed hurry, it was important, and if they were not lit up blue and red with screaming sirens, the drivers would just as soon their vehicles were not noticed. The last location these three yet-to-be-identified vehicles were seen was six miles from the fire and going like hell - without lights or sirens. It now appears these vehicles were dispatched to evacuate the sheriff from Camp Sierra and avoid her association to a pot grow and fire. They may have also participated in the removal of the marijuana grow evidence and attempted to put the infant fire out.

Information requested through a public records search for Fresno Sheriff's vehicles and radio GPS information for September 4th, 2020, was not produced. They said this information did not exist. They gave me names and badge numbers of those responding after 6:33 PM. These were the law enforcement that arrived on scene well after the 6:33 PM call-out, but requests for records of other vehicles in the area were never honored. These later arriving deputies, along with their patrol sergeant, could have cleared the marijuana grow area if requested to do so by U.S. Forest Service that evening, thus allowing a prompt engagement of the fire by bulldozer and hand crews. They were not asked by the USFS Incident Commander to do this. I believe sheriff's employees were already there and ultimately removed evidence of the marijuana grow - an unsanctioned, covert eradication that became the basis for the coverup of the true cause.

During the twenty-minute period before the fire response began, multiple attempts at getting a response to the fire were attempted but failed. The initial sheriff's radio traffic did not elicit a response, nor did the call to the sheriff's non-emergency number. The three calls on two separate cell phones to the sheriff's 911 dispatch center received busy signals, and finally, a landline call gets through to 911 only to be mishandled by the sheriff's department. It would be another 15 minutes from the time of this 911 call until a response is dispatched. Was this because of incompetence, neglect, or design? This amount of time, coupled with that lost between the first documented sighting of smoke (5:30 PM) and the first report of the fire which initiated a response (6:30 PM), wasted one hour of firefighting time which would have been more than enough time to contain it. Add this to the hour that a full response was staged (7:30 PM) and awaiting orders to approach the fire (8:35 PM) and you have over two hours of lost time. Remember how important it is to get to a fire quickly.

Time is of the essence when it comes to fire. The micro-second flash of a match, lighter, or lightning bolt is all it takes to produce a fire of hundreds of thousands of acres or more, and one that can burn for months. Creek Fire: 380,000 acres, 3 months and 20 days to containment. August Complex: 1 million acres, 2 months and 26 days to containment. Engaging the fire as close to its time of conception is the most important of the many variables that go into extinguishment. Every big fire was once a small fire that could have been put out under ideal circumstances. The Creek Fire remained a small, accessible fire for two hours before it exploded. Why was the Creek Fire not extinguished promptly?

Chapter 6

The Timeline

The following where times are stated was taken from radio recordings of the USFS Sierra Net Dispatch Channel:

6:33 PM: The person asked to verify the fire does so and walks into the USFS barracks to report the fire to Patrol 34. Within the next minutes, two helicopters, an air tanker, two bulldozers, eight engines, two hand crews, and numerous individuals are summoned by dispatch and are responding.

So ends the secrecy of the sheriff's fire and begins the uncertainty of the Forest Service's fire.

Why was it not CAL FIRE's fire? If it had been responded to by CAL FIRE alone it would have been put out in time to have most firefighters home by bedtime. If the USFS Engine 42 (E42) crew of five had extinguished the snag and scratched in a couple hundred feet of hand line before leaving, the less than one acre fire would have very likely reached containment that evening. An acre is 208 feet to a side; half this is not much of a fire. By this time the next evening, the fire was making its historic 15-mile run through the next county.

~6:40 PM: First photo taken from Big Creek Point showing only smoke. No flames visible from fire area.

6:56 PM: Entrance to road leading to fire flagged with neon ribbon by volunteer firefighter. First flame seen with fire moving up small ridge per USFS Silviculturist watching from Big Creek Point.

No mention of fire in snag.

6:59 PM: Air Attack 15 on scene and reports fire in creek bottom on lower third of slope and backing toward creek.

Spotter planes, in this case Air Attack 15 (AA15), are small planes without the capability to carry water or retardant. Their purpose is to manage the fire and direct the firefighting of both the ground and air assets. It is at this time that a reported sighting by the pilot of people running from the fire was allegedly heard by someone listening to a radio scanner. This was given to me secondhand with added information suggesting the pilot also stated that he believed the people had been attempting to put the fire out. I could not substantiate this information, although it fits quite nicely with the supposition that there were people in the area with knowledge of the fire that wanted it out and wanted to avoid being identified.

No mention of fire in snag.

7:05 PM: USFS Battalion Chief 33 (BC33), who is in transit, announces that the fire should not be approached from above.

7:08 PM: E42 is on scene and following flagged road toward fire, approaching the fire from above already. Fire is creeping per USFS Silviculturist stationed at Big Creek Point.

No mention of fire in snag.

7:10 PM: Specific directions to BC33, who is in transit, are given by USFS Region 5 Silviculturist on how to access 4WD road to fire. It is explained as an abandoned road reopened for a timber sale. It is described as okay for 4WD pickup trucks, but not engines. It is at this time that E42 breaks into the radio conversation.

7:12 PM: First retardant drop made by a British Aerospace 146. First CAL FIRE equipment on scene, Engine 4352 out of Shaver Lake Station.

The British Aerospace 146 is capable of delivering 3,000 gallons of fire retardant. This is generally released at one time, but on the Creek Fire, the first drop was half the load, a standard option with many air tankers. This half load was dropped on the eastern edge of the fire from an altitude above ground of over 800 feet. It vaporized long before reaching the forest canopy. My search of this area produced only a hazing of pink retardant on light-colored granite rock with the occasional larger droplet. Nowhere was there a coating sufficient to inhibit combustion. The discharged retardant produced absolutely no effect on the spread of the fire through the leaves and needles on the forest floor.

7:13 PM: E42 already on described road and states that it is not that bad, about as wide as the engine, and will turn around when possible. BC33 asks the E42 engineer to 'put eyes on the fire' before returning.

7:16 PM: Radio recording states fire is now burning snags (dead standing trees).

This is the first acknowledgement of the large snag burning, which was later determined (incorrectly) to be the source of the fire by USFS investigators. Photo from the 7:13 PM retardant drop shows no flame in snag.

Photo from the 7:21 PM retardant drop shows the snag fully engulfed. This fits with the evidence that the fire started below the snag and burned up to it.

7:18 PM: USFS Fresno 4 on scene.

These hand crews are generally comprised of about a dozen firefighters, sometimes as many as 20. They are often referred to as Bluecard Crews and designated by the USFS as a Type II team. These individuals are led by qualified USFS Firefighters, but most only have training in basic firefighting skills. They often speak limited English and are never to be in proximity to active fire. Their job is to follow the initial response well after active fire has been suppressed, cleaning up firelines and extinguishing smoldering areas within the containment line. A Freedom of Information Act request (FOIA) showed, on the daily payroll accounting, two Type II hand crew teams working that first evening: Fresno 4, and later to arrive, Fresno 3. No Type I team, which is trained for initial response, worked that first evening of the fire. As the E42 crew was put in harm's way by their unqualified boss, so were the two Type II crews. Neither should have been in proximity of the uncontrolled fire. While E42 safely drove away from the fire at 7:30 PM, Fresno 3 and 4 barely escaped the fire on foot when it exploded at 9:00 PM.

7:21 PM: Last half of retardant load dropped by air tanker.

A time-stamped video shows this last retardant drop from the approach of the air tanker to the target zone and until several seconds after the release of the retardant. A freeze-frame from this video allowed measurements to be taken and compared to the height of the burning snag (measured at 94’ 8” by USFS Fire Investigators)

Last of two retardant drops made by this plane. This 7:18 PM drop was made over three times the effective height above target and proved worthless in hindering ground level combustion. Credit: Toby Wait

and the length of the plane (86' to 100' depending on model). These measurements suggest the drop was made at least 900 feet above ground level. Recommendations for this plane suggest a full load should be deployed from ~200 feet above target and a half load from ~150 feet. This distance from ground or tree canopy produces the most effective coverage as the forward motion of the retardant has ceased and gravity has taken over, laying the liquid down as if it were a moderate, windless rain. USFS guidelines generally specify a 250-foot release distance above target for safety. Both retardant drops the evening of the fire were uncharacteristically high and absolutely ineffective, as were the two helicopter drops of water. Why?

The helicopter that dropped the two extremely ineffective loads of water on the nascent fire was a privately contracted ship stationed in Fresno. This ship supports both CAL FIRE and the USFS depending on the location of the fire. It was requested that evening by USFS Patrolman 34. The ship dropped twice with around 1,000 gallons released each time before returning to Fresno. Normally this much water on that small of a fire would have washed it off the map. The heavy lift helicopter used that evening was a Sikorsky 54, which uses a suction system to fill its reservoir. This limits the proximity of the helicopter to the fire as water must be released above the forest canopy. Releasing the water from a bucket suspended from the end of a 200-foot cable, where few, if any of the trees in the area reach more than 100 feet tall, could have put the water down accurately and effectively. This is what would have happened if a Chinook was available to respond. Witnesses say, as did the USFS employee over the recorded radio transmissions, all aerial drops by both the fixed wing and helicopter were ineffective. Those watching commented on the great distance from the ground all drops were made. There was at least one other helicopter that responded briefly to the fire from its base near Oakhurst (USFS 534). This ship flew over the fire initially, then proceeded to a nearby heliport where it landed. With information taken from flight tracker, this ship never approached the fire again and headed back to base directly from Ely Heliport. It was a line-and-bucket ship which could have hit the fire effectively.

No reason has been given as to why it did not do at least one drop.

Why does it appear pilots were uncomfortable approaching close to the fire? Was there knowledge of a marijuana grow there and gunfire from the ground a possibility, or was there an assumption of this possibility? Blame it on lightning and these questions become irrelevant. A retired air tanker pilot told me he often knew he would be flying over a grow and had the option not to fly. He said pilots frequently suspected a fire was grow-caused because of locality and proximity to water. Was this why the air attack on the Creek Fire was limited and ineffective? Was information relayed by the AA15 pilot to approaching aircraft of this possibility?

7:21 PM: USFS Battalion Chief 33, upon arrival at scene, takes the Incident Command (IC).

He would keep this designation until replaced by a CAL FIRE Chief when the fire began threatening private property near Shaver Lake in the early morning hours of September 5th.

Up until that time, CAL FIRE was not involved in fighting the fire. Initially responding CAL FIRE engines were ordered to stage within the community of Camp Sierra for possible structure protection. This was the decision of the USFS Incident Commander. It was brought to my attention that there was a heated discussion as to which agency had authority. USFS argued that it was their fire because it was burning on Sierra National Forest land leased by private individuals. CAL FIRE argued that the private structures on the leased land were threatened and that made the fire engagement theirs. It appears that if the CAL FIRE Battalion Chief had arrived on scene before the USFS Battalion Chief, he would have taken the IC position. What difference would this have made? The two agencies fight fire quite differently. CAL FIRE's approach is much more aggressive. It needs to be. Fires on private lands inevitably threaten structures immediately and must have as much firefighting strength thrown at them as quickly as possible to limit this loss. The USFS strategy is to manage fires. They will not rush decisions. They will analyze options with safety always foremost in their planning.

Time is of the essence to a CAL FIRE engagement. Planning is of the essence to a USFS engagement. My opinion only, but years in the making: CAL FIRE would make a snap decision to break protocol and save structures, placing them in potential physical risk and possible rule violation jeopardy. The USFS would take their time strategizing the engagement and analyzing risk, seeking supervisorial approval on anything that might break protocol before approaching the fire. Being that the Creek Fire threatened both structures and forest, both attitudes came into play. An example of each of these ingrained natures was present on the Creek Fire that evening. CAL FIRE suggested laying hose the 1,400 feet from the asphalt road above to the fire below. I understand a comment was made that the fire would be out in thirty minutes if this were done. This offer was not accepted by the USFS IC and the engines and battalion chief were directed to stage at Camp Sierra. At about this same time the USFS Silviculturist monitoring the fire from the Big Creek Point, over the radio, suggests the fire would be reevaluated the next day. One agency would jump on it right now - the other already expects the fire to be fought the next day. CAL FIRE fights fire. The USFS manages fire. The difference that evening in philosophies would have saved a great deal - likely 380,000 acres and 652 homes.

7:23 PM: Fire at less than one acre per aircraft AA15.

7:24 PM: Three aircraft released from the fire and returning to base: Air Tanker 160 and Helicopter 25M to Fresno, and Helicopter 534 to Oakhurst. AA15 still on scene.

The fixed wing aircraft, Tanker 160, would not have had time to reload and return a second load of retardant to the fire before dark, thus justifying its release. At this time there was over one-half hour of sunlight left and the heavy helicopter, 25M, could have made several more drops. This ship had filled twice from Shaver Lake at a four-mile distance from the fire and would have had time for at least five more drops from this source before darkness. If Balsam Forebay was used to access water, 2 1/2 miles distant, 8 to 10 more drops could have been accomplished before dark. If the airship was a Chinook bucket craft and Big Creek Powerhouse #1 afterbay was used like it was in the '94 Big Creek Fire, the 1 1/2 miles distance to the fire would have allowed a dozen or more drops by the 7:52 pm sunset of September 4th, 2020. This source was not an option for the Sikorsky as it would have had to set down within feet of the water to fill. A Chinook with bucket could have filled like it did in '94 and easily extinguished the infant Creek Fire. Why Helicopter 534 flew over the fire but never returned to fight the fire - instead returned to quarters - is also odd. None of these questionable actions have been addressed. The only reason I can deduce, and the only one suggested by others for all of these questionable actions, is a threat to the airborne assets and crew. Protocol was being followed in relation to a known or assumed criminal threat.

7:25 PM: USFS Engineer 42 radios the IC stating they are on their way out from the fire. IC tells him to switch to the tactical channel.

Tactical channels are not recorded by the USFS. They are intended to be used in the area of a fire between individuals relaying information not pertinent to dispatch or logistics, thus freeing the dispatch channel from on-scene radio traffic. Although not recorded, several people heard the conversation between Engineer 42 and the IC over scanners and USFS radios. All their descriptions of the transmission were similar. Paraphrased, 'We have a problem down here. There's evidence of a grow. There might be booby traps and chemicals.' At this point the Sierra Net Dispatch Channel goes nearly silent. Most all responding parties are on scene and need no further direction from dispatch. Those not yet on scene are heard from occasionally but are already aware of their orders. At this point all USFS employees on scene are communicating by way of tactical channels. The IC is the occasional exception. He is still in contact by way of the dispatch channel with Fresno 3 hand crew and Dozer 2, who are still in transit.

With radio traffic nearly ceasing after reported marijuana grow evidence was found, cell phone calls from the IC begin. From 5:00 PM to 8:04 PM, the IC had no incoming or outgoing calls. At 8:04 PM he makes his first phone call in hours, then makes or receives EIGHT more before the fire blows up at 8:59 PM. He then makes or receives a dozen calls between 10:00 PM and 1:00 AM. Numbers called by both the IC and the High Sierra District Ranger were redacted and a follow-up FOIA request for those numbers was denied.

7:32 PM: The hourly fire report. Fire still less than one acre and backing toward Big Creek (this would be the creek itself and not the town).

7:32 PM: CAL FIRE Engine 4381 on scene and directed by USFS IC to stage at Camp Sierra.

All CAL FIRE and local volunteer assets, along with Fresno Sheriff's Deputies, were staged at the community of Camp Sierra, while USFS assets were staged at the access point to the fire, which was 1/2 mile from Camp Sierra and about 3/8th of a mile by road from the fire. No USFS law enforcement responded that evening. An honored FOIA request showed them all off duty.

7:47 PM: AA15 still on scene.

7:48 PM: CAL FIRE Engines 4382, and shortly after, Battalion Chief 4312 stage at Camp Sierra.

7:52 PM: USFS Dozer 2 is on scene without swamper (an assistant that generally drives the support vehicle and does on-ground reconnaissance).

7:58 PM: Fresno 4 assists Dozer 2 in unloading the bulldozer from the trailer at USFS staging area.

8:05 PM: AA15 still on scene.

8:09 PM: Fresno 3 hand crew on scene.

At this point all responding assets are on scene and staging at Camp Sierra or at the access road to the fire. No on-ground firefighters have approached the fire other than E42 at 7:20 PM. This would remain the case for the next 80 minutes. From 7:20 PM to about 8:40 PM, no firefighting took place. Before that, four impotent air drops and an ill-advised approach to the edge of the fire by E42 was the extent of engagement.

8:32 PM: The Incident Command is asked for a fire update. I will quote him from the radio recordings. IC: "Fire is holding at a little over one acre. No growth currently." Dispatcher repeats statement, states the time (as is protocol) then asks, "Are you available for a phone call?" IC responds, "Affirmative." Dispatcher: "Can you place call to office?" IC: "Affirmative."

A FOIA request of phone records for the IC show the call to 'the office' was made by IC at 8:33 PM. This call was made to Fresno. The called number and party were not divulged in original FOIA (redacted). This call could have been to any of the supervisors in the fire hierarchy, of which there are more than half a dozen.

In the minutes after this call two hand crews and a dozer began approaching the fire by way of a well-maintained power line access road, the same one E42 had driven 90 minutes earlier. About halfway to the fire a hard right turn put the hand crews and dozer on a hundred-year-old communication line road that had been opened up for access to a proposed USFS timber sale earlier in the year. From this turn down to the fire, the road was very narrow, somewhat steep, and extremely dusty. Four-wheel drive would be necessary to exit the fire area.

Along this road, hand crews cut roadside brush to produce a safe-from-burnover escape route with the dozer following. From near the end of the road the dozer and one hand crew turned left, away from the fire, and began working. This was the point where E42 had turned around earlier in the evening as evidenced by the loss of considerable water which had sloshed out of the engine's open-topped tank. At this point they were about 200 feet from the end of the passable road and 300 feet from the fire. Also from this point, the other hand crew turned right and began cutting a trail to the southeast corner of the fire. This P line, short for pioneer line, dropped off a steep ravine to a small spring-fed creek. It was here that Fresno 4 found evidence of a nearby grow. No plants were found, but lightly covered drip line ran from the creek to the top of the hill they had just descended. A fuel can, a car battery set up to charge cell phones, black buckets, and food trash were also seen. The drip line headed in the direction of the working dozer.

From both the vantage of Big Creek Point and Camp Sierra, witnesses both saw and heard the dozer and hand crews. It was dark by now, but the chainsaws could be heard pruning brush and limbs, and headlights of the sawyers could be seen as they led the way. Behind them came the brush handlers, pulling brush and limbs away from the escape route, and behind them, the well-lighted dozer clattered along widening the road as needed. These same witnesses watched as the dozer made the left hand turn away from the fire, taking a group of head-lighted bodies with it. Only the glow of the fire could be seen from Camp Sierra at 8:45 PM as the flame was below the line of sight. Above and to the west of the snag, which had quit burning by now, the dozer could be clearly seen working across a contour directly away from the fire. In all appearance a fireline was being pushed in. In reality, buried buckets containing, or having contained, marijuana plants were being pulled from the ground, or had already been removed before the dozer arrived, and the area was then being torn up by the tracks of the dozer to hide this evidence. No fireline existed along this contour in mid-January of 2021 when we first accessed the point of origin area.

What we did find were dozer tracks over an area 500 feet long and 20 feet wide where the dozer had made a pass out, constantly turning right and left while moving forward, then returning beside the first tracks to do the same. This was recognized by those of us familiar with forest practices as the standard maneuver to crush brush and disturb ground cover in preparation for tree planting or seed broadcasting. Or to hide something from sight, like the bucket impressions from an illegal marijuana grow.

Months later we would acquire a video showing buried 5-gallon buckets with drip line from a nearby creek running to them. This grow of six buckets, video of which was taken the week prior to the Creek Fire, was lacking plants. It was obvious by the growth of natural vegetation within the buckets that these had not been part of a recent grow, but it gave us a clue to reinvestigate the dozer trail. A retired cop with infinite patience found two bucket impressions under the disturbed ground of the dozer tracks. It appeared they were not only run over by the dozer, but also carefully filled in by hand and camouflaged. A third impression would be found later by the lead USFS fire investigator. Initially we were looking for evidence of a larger, cartel-type grow. Upon seeing the video of the six-bucket grow on the adjoining creek, our idea of the grower's profile changed. It was no longer a big money grow of tens or hundreds of plants overseen by a couple of cartel hostages, but a few plants tended by a local. Even though they were being grown on national forest land, which is illegal for any number of plants, by growing only six, any serious charges would be avoided if the growers were caught. Six plants are the maximum allowable for personal use; at seven plants you become a commercial grower and drug dealer.

8:59 PM: The fire blows up. The dozer operator is heard on the recorded radio traffic frantically announcing the need for him and the hand crews to leave the area due to extreme fire behavior. The Fresno 4 hand crew must also escape the fire area, and like the dozer, left without constructing any fireline.

This ends any opportunity to control the fire.
Time has run out.

Over the next three hours the road out of Big Creek is threatened, and all traffic except fire responders is redirected to Huntington Lake in order to escape Camp Sierra and Big Creek. Shortly after midnight the road becomes too dangerous for fire equipment and is shut down to all traffic. At 12:28 AM a 911 call is received asking if the residents of the private railroad museum and home located on the SJ&E Railroad Grade just west of Huntington Lake Road should evacuate. It seems this structure had been overlooked by both the USFS and Fresno Sheriff's Department in their evacuation notices. The residents escape shortly before this historic structure and its artifacts, dating to the time of the SJ&E Railroad, are destroyed.

The Forest Service dozer, having left the point of origin area because of fire behavior, moves to another area and breaks down while building fireline below The Point. This necessitates CAL FIRE Dozer 4245 being requested for the second time in eight hours. This dozer, on transport, was turned around at about 8:30 PM the previous evening in Auberry, less than 25 miles from the fire, and returned to quarters. While the USFS IC is requesting a CAL FIRE dozer, he is releasing four staged CAL FIRE engines, a volunteer engine, and a water tender back to quarters. None of this equipment left Camp Sierra since being staged there around 7:30 PM and none of this equipment discharged any water while there. Ironically, they cannot return the way they came to the fire - the road to Shaver Lake is now burning on both sides and no vehicles can pass. Five engines and a water tender must return by way of the back road to Big Creek, around Huntington Lake, and through the resort town of Lakeshore to Shaver Lake, adding an extra 25 miles to their return to stations.

3:00 AM: At about this time, as the fire burns into private land, CAL FIRE takes over management. The fire has now moved two miles from its point of origin and has set itself up to make threatening runs in several directions during the heat and winds of the coming day. It is poised to do what no fire has done in seventy-three years: threaten the town of Shaver Lake.

In the summer of 1947, a fire started at the Bretz Brothers lumber mill in Blue Canyon and quickly burned up-slope toward Shaver Lake. The town quickly prepared to evacuate, but after crossing Dinkey Creek Road, the fire lost its momentum. Here it encountered flat ground with multiple meadows and the nearby ridges were sparsely covered with brush and timber with numerous granite outcroppings. It was not a good fire environment, and the fire burned itself out on the far side of the lake from town, not having the requisite fuels needed to sustain an actively moving front. It lasted but an afternoon as a threat to the town and burned a meager 3,000 acres. The Creek Fire was already a different flaming serpent than the Bretz Mill Fire. Where the Bretz Mill Fire had only one well-fueled avenue to follow, and only one head to worry about, the Creek Fire would become a Medusa with multiple flaming fronts - all its potential pathways aimed at a different community.

Big Creek—Saturday, September 5th, 2020

Although the fire started less than one mile from the Big Creek Point and less than a half mile from the nearest cabins in Camp Sierra, the first night's burn took it directly away from both communities. Morning brought a short-lived optimism as the fire was still burning away from structures, climbing the eastern flank of Musick Mountain toward its fire lookout and antenna array. It was also moving downslope toward the San Joaquin River at its confluence with Big Creek. Burning very actively downhill during the night, assisted by the down-canyon winds, it was poised to be a problem in multiple directions come morning. It did not wait for sun-up.

The extreme heat moving into the San Joaquin Valley with forecast temperatures of over 100 degrees was already having an effect on the fire. The air from the valley floor was funneling toward the sailing mecca of Huntington Lake early, and the fire was responding under the steady breeze that would soon become a gale. It was obvious that Big Creek and Camp Sierra would quickly be facing a well-fueled wall of fire. At 4:00 AM the call to evacuate was sent out. While fire trucks responded one direction on the narrow mountain road, evacuees escaped in the opposite direction. Shortly, the town was abandoned except for a handful of residents who chose to stay and fight and a few engine crews that had mustered out of the surrounding area in hopes of saving Big Creek and Camp Sierra.

By sunrise the fire was coming up both sides of the creek and fire crews from several communities fought heroically to save the two communities throughout the day. They were successful in protecting most of the cabins in the summer hideaway of Camp Sierra, losing only four. Big Creek caught the focus of the funneled flames in the afternoon. There was not sufficient air support this time around. The flames came like they did in '94, this time threatening the houses along the north side of Point Road first. The engine crews attempting to protect these structures faced instant futility. When they gave up the fight, houses on the south side of Point Road had already been ignited by those burning on the north side. The wind was blowing fire horizontally across the road when the last engine escaped through the flames. A Fresno engine crew was trapped by downed power lines and had to retreat to the safety of the heliport.

Within minutes the fire moved into Jungletown and within minutes Jungletown was fully engulfed, along with numerous homes on the road leading to Huntington Lake. All told, 41 homes burned in that brief firestorm - nearly half the town - including the school house we lived in during the '94 fire and the home my family lived in before I was born.

View seen from Shaver Lake by several thousand recreationists the afternoon of September 5th. Within five hours Big Creek was burning and by nightfall nearly all of Huntington Lake area was aflame. Credit USFS via internet

Saturday Morning, September 5th, 2020 – Mammoth Pool Reservoir

That first morning the Creek Fire was positioned to make multiple runs. As it was moving up canyon toward Big Creek, it was also advancing northwest toward the rural areas outside of North Fork. The fuels it was encountering were very volatile being comprised of mostly dense brush and low growing oaks with scattered Ponderosa pine intermingled. Under oven-like conditions these fuels became explosive, and by mid-morning in the San Joaquin River Canyon, near its confluence with Big Creek, another run of the fire would take off, pushed by southeast winds crossing a sun-baked valley.

The fire had burned down the northwestern slope of Musick Mountain throughout the night and morning, approaching the steep river canyon before noon. Under normal fire conditions the canyon of the San Joaquin would be a major impediment to fire spread; it had been in the past, only being jumped by a fire in 1937. No one would be prepared for the fire behavior of this afternoon and evening; those spending the weekend up-river near Mammoth Pool were unaware that their plight would soon be broadcast internationally.

Fire activity increased the closer the fire approached the river under the influence of the rising temperatures and increased winds. Soon the speed of its forward progress consumed enough fuel to build a column of smoke several thousand feet high (over 50,000' by afternoon). The convection process feeding the flames with vast amounts of oxygen at its burning base increased the winds throughout the area. As long as there were abundant, dry fuels to burn, this convection system would continue to build; the winds would continue to increase, and the temperatures would continue to rise until the fuels ran out. At this time, the smoke column would collapse under its own weight and spread fire in all directions.

When conditions are right to build a several-thousand-foot column of smoke, spotting begins to occur. Partially combusted fuels such as bark, limbs, and cones will be sucked up in the vacuum of the convection and will be transported within the column without com-

The column produced by the fire jumping the San Joaquin River into heavy, volatile fuels the afternoon of the 5th. This photo was taken from a plane flying from San Jose to Las Vegas and shows the smoke approaching 25,000 feet. Later that afternoon it would reach over 50,000 feet and produce lightning and ash-filled hail before collapsing in a firestorm over Mammoth Pool Reservoir and trapping over 200 people. Credit: Thalia Dockery

pletely burning. The interior of the column is nearly devoid of oxygen - it being consumed quickly near the ground - and is uniformly heated to temperatures nearing 1,800 degrees, sometimes higher. In physical terms, the column of smoke is a cylinder of turbulent mass several miles high and weighing thousands of tons. This weight, suspended by heat alone - much like a hot air balloon - will eventually lose its supporting energy upon cooling and collapse. Wind speeds will exceed 100 mph in this downdraft of falling debris and ash. Prior to collapse, unburned material will come into contact with the oxygen-rich air near the edge of the column and burst into flames.

If these burning firebrands are discharged from the column in proximity to the ground and land upon combustible material, another fire is often the result. This spotting makes fires exponentially more difficult to fight and much more dangerous. Instead of one, there will be several fires that can surround even the most experienced firefighters. On the front line it is extremely important to always have someone watching your back and have a safety zone identified and a plan to reach it in place.

This spotting distance is often over a mile from the main fire and is common in dry, windy conditions. On this day, the Creek Fire would double and triple that spotting distance. One eight-foot-long cedar limb fell to the road nearly five miles in advance of the fire front, totally charred and still smoking. It was firebrands like this, cast across the river canyon multiple times, that produced the perfect storm responsible for the formation of the massive smoke column which then became a pyrocumulonimbus cloud capable of producing lightning. The spot fires in advance of the main fire were then sucked back toward the fire front, contributing their smoke and heat to the building column. As many as a half dozen of these spot fires were seen ahead of the Creek Fire as it roared up the San Joaquin River Canyon.

A cloud of this nature occurs over nuclear detonations, volcanoes, and the largest forest fires. The heat of combustion releases any moisture from the fuel it is burning, and near-surface soil will have its dampness boiled away; even water bodies such as lakes and streams will contribute evaporative moisture to the updraft. This water vapor joins the smoke, ash, and embers in forming the column, and upon reaching the cooler temperatures of the higher atmosphere it will begin condensing into thunder clouds which occasionally produce lightning, which this cloud did. The pyrocumulonimbus cloud that formed over the Creek Fire reached 52,000 to 55,000 feet that day and is possibly the highest reaching wildfire column ever recorded in North America. The ash that was lofted into these fire-born thunder clouds mixed with the condensing moisture to produce hail of an unnatural dirty-gray color that fell to the ground in the Huntington Lake area.

This incredible event would spell horror for nearly 1,000 people recreating in the distant mountains, but first it would surprise several nearby homeowners in the Kinsman Flat area with evacuation calls stating: IMMEDIATELY EVACUATE OR FACE IMMINENT DEATH; DO NOT GATHER PERSONAL BELONGINGS; LEAVE NOW. The warnings were heeded and everyone survived. Anyone familiar with this type of fire with its size and intensity would be amazed that no one perished here or elsewhere on the Creek Fire - a fact that weighs heavily on the outcome of the fire investigation. The road these residents escaped on, Minarets Road, would soon be fully engulfed for miles by flames.

Aspen Fire, June 28, 2013

In late June of 2013, an afternoon thunderstorm formed in the Kaiser Wilderness area just north of Huntington Lake. This is quite common throughout the Central Sierras during the summer months, produced by a flow of moist air streaming from the south into the Four Corners region of the Southwest. A high-pressure ridge often sits over the confluence of the boundaries of AZ, NM, CO, and UT from June to September. This monsoonal flow, as it is referred to in weather jargon, is air pulled from the Gulf of California and, to a lesser degree, from the Gulf of America by the clockwise rotation of this stationary high-pressure system. Normally this extremely moist air stays to the southeast of California, but when the consistently west-to-east flow of the summer jet-stream winds shift north, it allows this moisture to cross the crest of the Sierras and produce thunderstorms in the mountains and foothills of Central California.

This weather pattern produced the lightning that started the 2013 Aspen Fire. Aspen Creek, for which the fire was named, is a minimal tributary of the San Joaquin River, existing for a short distance from the meadow of the same name to its precipitous cascade into the river four miles downstream from Mammoth Pool Reservoir. The Aspen Fire, not burning under extreme fire conditions, consumed 23,000 acres of brush, oak, and lower elevation scattered timber in the river canyon below the reservoir before spreading into heavier timber as it burned out of the river gorge. This fire consumed all combustibles within the canyon clear to Mammoth Pool Dam, leaving trees and heavy brush stripped of their foliage and the soil surface lifeless.

This was not the case below the soil's surface. Although the complete burning of the pine trees ended their life, the roots of the brush and oak survived the burn and would quickly sprout new growth - often within a month - and would soon become a nearly solid ground cover. Only under the most intense burns are roots of brush and oak totally destroyed. The nature of these plants has adapted them for survival in fire-prone environments. They become the pioneer species (the first to grow back after a fire) taking advantage of the nutrients released from the combusted organic material, increased soil moisture from lack of plants, and the abundant sunlight no longer filtered out by the over-story of conifers and established oaks. This opportunity contributes to incredible initial growth that will continue for over a decade before moderating. By this time, nutrients, moisture, and sunlight are becoming limited once again and a typical long-term growth pattern will be reestablished, but only after quickly contributing tens of tons per acre of fuel to the next fire. The Aspen Fire will become significant when the Creek Fire reaches its fire scar because of this process of nature.

French Fire, July 27, 2014

While the Aspen Fire burned on the cooler northern exposure of the San Joaquin River Canyon earlier in the fire season, the French Fire, caused by an escaped campfire, burned the more fire-friendly southern exposure. This, coupled with the later start date and drought conditions, made for a more intense fire. This southerly-facing aspect produced very good growing conditions for brush and oak within the canyon and hastened the growth of brush beneath the timber on the less steep slopes to the north. This brush/timber transition, much like the area around Big Creek, makes for one of the worst fire fuel scenarios that exists. The varying drought conditions over the previous few years had made trees, primarily Ponderosa and sugar pine, susceptible to bark beetle infestations. Over the next several years an estimated 163 million trees would die in the Sierra Nevada Mountains because of these beetles. The area of the French Fire, and to a lesser degree the Aspen Fire, was affected by this mass die-off. With the brush growth under the dead pine trees, drought conditions, and canyon winds, a widespread volatile fire situation was present. Put simply, fire was waiting to do what the lack of forest management, which includes logging, had done.

Like the Aspen Fire, the French Fire burned up canyon from its point of origin three miles below, and all the way to Mammoth Pool Reservoir and beyond. Unlike the Aspen Fire, the French Fire burned intensely; about half of the 14,000 acres consumed in the French Fire were described as having burned in this manner. The dead and dying trees affected by the beetles combusted easily, leaving blackened stems devoid of needles and limbs standing over soil that was now nearly sterile from the heat of the burn. Most of these trees had either died recently or were in the process of dying from beetle damage and still showed green somewhere in their crowns. These trees still held considerable moisture beneath their bark thus protecting the wood from damage. Nearly all these trees would be of value to the logging and lumber industries, even in their charred state, if harvested promptly. After three years, all but the largest would be of no value for lumber. Of the acreage burned, less than 25% was allowed to be logged. Thousands of trees were not only wasted as lumber, but the opportunity for local employment to harvest them was squandered. Besides these political and economical issues, the contribution of this flammable mass left to future fires was ignored; ignorance that would ultimately threaten human life.

Little infrastructure was damaged in either of the fires. No homes or lives were lost. Where the fires burned and how they burned limited loss of wildlife. The biggest loss was to timber. Prior to their burning, the beetle-killed pine trees that helped fuel the fire were being salvaged by logging companies along the road corridors for safety reasons. This accelerated after the fires to include burned trees, but only where they posed a threat to people driving the roads or using recreational areas such as campgrounds and picnic areas. Beyond a couple hundred feet from the road, no trees could be cut, whether bug or fire-killed.

This directive was the result of years of incrementally protecting more and more aspects of the forest. No order of life within the biomes of the forest is ignored if it can be identified as rare or endangered - and all will be protected. This process, whether by intent or not, has curtailed logging to the point of being economically unfeasible. For over a hundred years the removal of timber by private industry was the primary economy of most mountain communities. During this period, few fires, if any, approached the size that are regularly seen now in the Sierra Nevada Mountains. After the Gold Rush, logging replaced natural fire and acres logged replaced acres burned; this was the case until the 1970s when a hippie's view of forest management took over the U.S. Forest Service. In 1980 there were nine lumber mills sawing timber from the Sierra National Forest; in 2020 only one of these remained. There is more to blame beyond global warming for the current fire climate.

Minarets Road - Saturday Afternoon, September 5th, 2020

While the Creek Fire was threatening private property near Kinsman Flat with immediate evacuation and forcing imminent evacuation for areas toward the towns of North Fork and Bass Lake, it was also beginning its assault on the San Joaquin River Canyon toward Mammoth Pool Reservoir.

Mammoth Pool sits 40 crooked, mountainous miles from North Fork by way of Minarets Road. The hour and a half that it takes to navigate this trip lends to the nature of this scenic route. Several hundred people, possibly a thousand, were in the vicinity of the reservoir for the Labor Day Weekend. Nearly all of these people drove this route to their campsites. Few knew of the dirt roads intersecting Minarets Road that offered alternate routes to the lake, and those that did know had no intention of adding another hour to their trip by using them.

Near Kinsman Flat the San Joaquin River and Minarets Road are less than one mile apart. From this point to Mammoth Pool by road is 25 miles; from this point to Mammoth Pool by San Joaquin River Canyon is 7 miles. Both sides of those seven miles of river canyon had burned in the previous decade by the Aspen and French Fires. The ash-nourished regrowth from these fires had produced brush as tall as ten feet in places. The extremely dry dead oak and pine snags added greatly to the fuel load.

Once the fire jumped the river canyon, its path to Mammoth Pool was set. Burning on both sides of the river and pushed by up-canyon winds, the fire moved more quickly upstream than either the Aspen or French Fires. As it burned up canyon it also spread out of the canyon to the north and south, quickly engulfing Minarets Road and isolating the campers who had used it to access the forest and lake above. Over the next few hours, those camping in the Mammoth Pool area would become aware of the threat the fire posed by the growing smoke column and falling ash, but they were ignorant of the variables that were intensifying the inferno below them.

Whether by a previous plan or the threat the fire might pose, multiple vehicles started the trip down Minarets Road toward North Fork late in the morning. Initially the path of the road took them away from the lake and river which distanced them from the oncoming fire, but soon the road swung back to the south, and through their front windows an ominous plume of smoke was rising ever higher into the sky. If not for a decision made by a local North Fork resident and his wife to drive toward Mammoth Pool on Minarets Road to view the fire across the canyon, those driving down from the reservoir may not have survived.

Having an interest in one of the isolated holdings of private land within the national forest a few miles west of the reservoir and having worked to retirement in the logging and lumber business in the area, this couple had more than a passing interest in where the fire was and what it might do. The last thing on their minds was the thought they might be endangered by the fire. After all, it was on the other side of the river - a barrier not breached by fire in over eighty

What was seen by North Fork residents, Jon and Judi Norby, across the San Joaquin River Canyon just before the fire jumped to their side and began building a massive fire column of smoke, ash, debris, and moisture. This river canyon had prevented many fires from spreading until the Creek Fire. The last fire to jump the river was a small blaze in 1937 that was quickly controlled. Credit: Jon Norby

years - and still quite some distance away. With the knowledge that neither the Aspen nor the French Fire had jumped this barrier, they drove to where the best view would be found.

The fire was burning moderately when they embarked and the dozen air miles between them and the fire made it seem distant and unthreatening. As they swung around Castle Peak above Redinger Lake the fire came into full view. Now less than a half dozen miles from the western edge of the fire and looking directly into the flames, they saw the fire's potential become frighteningly more ominous. Any uphill run from here would put it into the canyon below Shaver Lake Dam and any lateral movement would put it just over the hill from the town of Shaver Lake. With its current direction it would soon be at the door to Jose Basin, directly below the town of Shaver Lake. Fire behavior was quickly getting worse and the fire was on the move. Would this become my father's worst nightmare? Would The Basin burn? Time would tell, but for now the North Fork residents' focus was on the river canyon.

At Kinsman Flat the local couple stopped to watch, comfortable with the fire burning on the opposite side of the river. The fire's head was now advancing under the steep, dry, westerly slope of Musick Mountain, having burned the moister northern exposure quite efficiently the night before. It was inevitable it would soon explode upslope, but first the hot winds out of the southwest would take charge and move it quickly toward the river and the couple's point of observation.

It was just ahead of lunchtime, the intensity of the fire increasing in the building wind, when smoke was detected below them. Any flames were out of view from their vantage point, but the smoke suggested that the fire was now burning on their side of the river canyon. A passing power company employee verified this, having heard over his radio that the fire had just jumped the river. The fire now had another head, at least its fifth, that would need to be severed to halt destruction. This would be impossible under these fire conditions in this inaccessible canyon.

The fire spread so quickly that to return in the direction they had come was not a sensible option to these lifetime residents of the area. They knew this country well and would travel toward Mammoth Pool and take the back way home. Through the open window of the truck the heat of the approaching flames could already be felt. Within a couple of minutes, the flames could be seen in their rearview mirror. In a few more minutes the sun was blotted from the sky and darkness began to spread over the whole landscape, muting the natural colors in a strange silvery tone. They felt as if the fire was gaining on them as they drove toward Mammoth Pool.

The fire jumping the San Joaquin River started the perfect storm which produced the tallest pyrocumulonimbus cloud in U.S. history, possibly as high as 52,000 feet. It also contributed to the fifteen-mile spread of flames over a twenty-four-hour period, one of the greatest distances traveled by a fire through forested land ever recorded,

Fire burning Kinsman Flat just minutes after jumping the river. Note the strong ground level wind feeding the fire's base before being directed upward into the building column. Also note the darkness under the building column which gave a metallic gray hue to the natural colors. Credit: Jon Norby

most of this distance in the first eight hours. Three tornadoes with estimated winds of 100 mph or more were investigated in the Creek Fire - two here, and one near Huntington Lake. It was at this time that the private residences in the Kinsman Flat area were given minutes to escape, some with their homes burning as they were leaving their driveways.

The first people met on Minarets Road, within minutes of the North Fork couple leaving their vantage point and heading toward Mammoth Pool, were a pair of young men in an older pickup truck. The driver of this truck was told of the situation but had only enough fuel to reach North Fork by this route and chose to proceed. He and a couple of drivers behind him, evacuating horses in trailers, braved the roadside flames and made it through. Further up the Minarets Road, more than a dozen other vehicles were encountered. All of these heeded the warning of an impassable road ahead of them and joined a convoy that headed back toward Mammoth Pool, led by the local couple. By this time burning embers were falling over a large area and spot fires were leap-frogging well in advance of the main fire front. The Minarets Road was burning on both sides for multiple miles behind the snaking convoy.

The escaping group had to negotiate the winding road at a safe speed and be ready for any vehicles coming toward them which would be attempting to evacuate the oncoming fire in a hurry. While the drivers wound their way toward Mammoth Pool by way of the road, the fire was taking a wind-driven beeline to the reservoir up the San Joaquin River Canyon. Any mishap here could allow the fire to overtake them from the canyon below. Soon they were to the Mile High Curve where the road hangs on the lip of the canyon nearly 2,500 cliff-like feet above the river. From here the Minarets Road leaves the canyon, and with the worst of the threat behind them, the fleeing campers made their way to the Grizzly Meadow turnoff and then to Beasore Road. Here the rescuing couple directed them toward Bass Lake and safety.

In all likelihood, the decision of the local couple to travel toward Mammoth Pool to view the fire saved lives. The group that was turned around was about five miles from where the immediate evacuation of Kinsman Flat was taking place; the fifteen minutes it would have taken them to travel this distance would have put them in the midst of the exploding fire and its long-range spotting. Chance played heavily in keeping anyone from dying in the fire, but wise decisions played a more important role. The next to get lucky with their lives would be those around the reservoir itself.

Mammoth Pool Reservoir - Saturday Afternoon and Evening, September 5th, 2020

Lunchtime at the lake was enjoyed with little thought of the fire miles away. Anyone concerned of a possible threat would have discovered from the broadcast news that the fire was approaching Shaver Lake from two separate directions and an evacuation of the town of Big Creek was complete. There was no mention of a concern to the Madera side of the river until the calls for immediate evacuation of Kinsman Flat came out. Within the hour thoughts changed as the threat down canyon became obvious; the quickly building cloud of smoke needed no official announcement of the danger. With the sun blocked by spreading smoke, and ash falling like snow, many made the choice to cut their weekend short. Fortunately for most, this decision was made in time to have access to the back way out of the area. Minarets Road had been blocked at Grizzly Meadow Road by this time and signs had been placed directing travelers toward Beasore Road and the towns of Bass Lake and Oakhurst. This option lasted but a very short time.

The massive column of smoke, now fully developed into a pyrocumulonimbus cloud producing lightning, was running out of fuel. The river canyon walls directly below Mammoth Pool Dam became nearly solid granite where vegetation became sparse. This divided the head of the fire and quickly denied it of the needed fuel to sustain the nine-mile-high convection system the smoke had become. In minutes the sky became a downpour of super-heated debris. Larger pieces burst into flames as they contacted the oxygen-rich air outside of the column. It was literally raining fire over a vast area under the collapsing column.

In a matter of minutes there was no escaping Mammoth Pool. Grizzly Meadow Road was now in flames, and those turning back only had a brief time to escape by heading past the Minarets Work Center to Beasore Road at the pack station. The road dropping down into the lake was quickly blocked by fire, trapping those that had procrastinated. The fire encircled Mammoth Pool, chasing out campers from both designated campgrounds and scattered primitive sites around the dam. By evening over 200 people were trapped in the lake bottom around the boat launch area. The smoke made breathing difficult and most covered their faces with wet cloth. Some laid under wet blankets to escape the heat. The store, residence, and multiple outbuildings burned. The residents and staff escaped to a nearby meadow and pond to outlast the flames.

Helicopters were dispatched from Fresno to rescue the victims but found visibility difficult, requiring navigation using night vision goggles. Flying blind into heavy smoke at midnight in mountainous terrain was a feat that all involved should be proud of. 214 people and 16 dogs were rescued successfully with the only injuries being those sustained prior to rescue. Within a couple of days those hospitalized were discharged. It would take weeks for some to recover their vehicles and what camping gear was not destroyed. The resort is still in the process of rebuilding five years later. Again, unbelievably, no lives were lost. This would become the mantra of the nearly four-month long fire: no lives lost.

A Thought

There was a very poorly conducted investigation of the cause of the Creek Fire. Was this intentional? Within ten days of the start of the fire, lightning was being suggested as the likely cause. Specific information I received in the first two days of the fire, from people I would trust with my life, pointed directly at the fire starting near a marijuana grow. I heard from multiple residents of what they heard over radios and scanners; the same message that was being passed around amongst law enforcement and volunteer firefighters. 'We have a problem down here. There's evidence of a grow. There might be booby traps and chemicals.'

I never considered lightning even a remote possibility, nor did most of the mountain community. Nothing, absolutely nothing, suggested this as the cause, yet it was chosen as a possibility by the investigators, or more likely, someone above them. If just one person had died as a result of the fire, even if by accident during the evacuation, or any one of a hundred other scenarios, the fire would have been investigated differently. Accountability and culpability would have been established. The still unanswered questions would have been addressed in court by participants under subpoena and oath. Lightning made it easy to sweep the burning of 652 homes and 380,000 acres under the rug. One death and the truth would have been exposed years ago. I would have seen to that. Without this motivation, individuals and agencies have not been held accountable - yet.

Huntington Lake - Saturday Evening, September 5th, 2020

The Big Creek drainage head of the Medusa, which the Creek Fire had now become, was now raging around Huntington Lake. It came north out of Big Creek just like the fire of '94 had that afternoon twenty-six years prior. This time, instead of sitting down when it reached the relatively flat ground at the top of the chimney above Big Creek Canyon like it did in '94, the fire pushed into the Jeffrey pine and red fir forest surrounding the lake.

The shores of Huntington Lake are home to hundreds of summer cabins, several church retreats, small resorts, campgrounds, and scout camps. Normal sailing winds would have been sufficient to threaten this narrow band of habitation with fire, but with the added energy of down-canyon combustion, the Creek Fire roared into the Huntington Basin. A remnant portion of the already small local volunteer fire department had been moving west to east through the clusters of cabins doing what they could to fireproof them throughout the day. They were now dealing with advancing flames and spot fires. They heroically fought fire with rakes, shovels, and garden hoses, house to house as the fire spread. They saved dozens of structures with their efforts that day, but the fight soon became futile and their lives became at stake.

Just like in the next drainage to the north - the San Joaquin River Canyon where Mammoth Pool sat - the perfect conditions for a firestorm came together that afternoon at Huntington Lake. Likely fed by the massive, turbulent column of smoke building five miles to

The fire as seen by firefighters as it left the Big Creek Canyon and began its assault on Huntington Lake. The energy put skyward by this mass of flame produced a thunderstorm with soot-filled hail and a tornado that uprooted several acres of trees, ultimately burning over 70 cabins and other buildings surrounding the lake. Credit Clovis Fire Department

the west on the San Joaquin, the winds became gale-like and erratic around the lake. Ground fuels were limited throughout the structures from normal cleanup and recent thinning had broken up the forest canopy which kept the fire initially on the ground. Those who had raked well themselves, and those that had been hastily visited by the volunteers earlier in the day, fared well; but several structures were still lost. Later, as the fire spread eastward, the winds got a hold of it in forested areas away from cabins, and in these pockets of dry fuel, the fire crowned.

The fuels, winds, and temperatures began producing another column of smoke and heat, and soon the fire began spotting. Those attempting to fight the fire had to escape the entrapment that was already happening. Spot fires began cutting off their single escape route. The flames moved into heavier fuels and sucked in the surrounding spot fires. Another pyrocumulonimbus cloud now formed with its associated lightning, and like to the west a tornado formed. This appears to be one of the strongest ever investigated, leaving acres of trees, some over 200 feet tall, laid on the ground in the pattern of the rotating wind. Wind speeds up to 125 mph were suggested. It was difficult to establish the speed of this wind because the scale used to measure this takes damage done as its basis. What size car was flipped over? What size tree was snapped? This was a mature red fir and Jeffrey pine forest that was leveled. One of these environments had likely never experienced a tornado before; it was new ground for the investigators to judge the force of the rotating wind.

From here the fire was free to roam. No person attempting to stop it would be safe and no amount of resources would help - if any were available. The extreme fire conditions that afternoon would soon be joined by the collapse of the fire column. When firefighters returned, they found a patchwork of destruction. Some large areas were untouched while others were completely incinerated. A total of 74 structures were lost, including the Huntington Lake Volunteer Firehouse, which burned while members fought fire nearby.

Shaver Lake - Sunday, September 6th, 2020

It was soon obvious the Creek Fire was uncontrollable on Sierra National Forest lands. Where fuels were sufficient, it burned easily; where fuels were heavy, it burned like no other fire in the area had ever burned. It burned hundreds of acres covered in hundreds of dead trees per acre, not one by one, but all at once - and the town of Shaver Lake was directly in the fire's path.

When the fire turned the corner under Musick Mountain the afternoon of the 5th, nothing stood in its way until it reached the private lands surrounding the town of Shaver Lake. Southern California Edison Company (SCE), the local electrical power producer, had acquired these timberlands early in the 1900s from several of the area's logging companies and their heirs. In 1927, in the midst of these holdings, a dam was completed which impounded the waters of Stevenson Creek. This lake covered the original town of Shaver, whose buildings and sawmill structure were set ablaze in the fall of 1926 in the clearing of the lake bottom for its imminent filling. The current town of Shaver Lake stretches from this dam west for several miles along Highway 168, interspersed with SCE lands. Many of these enclaves carry historic names like Shaver Point, Rock Haven, East Village and West Village, Musick Heights, Sunset Rock, Indian Rock, and Ockenden. To the west are other small, newer communities, one of which encompasses the apple orchard from which I viewed my first fire, now called Apple Ridge. Several 140-year-old apple trees still exist in the yards of homes there.

By mid-morning of September 6th, it had become obvious that a couple of thousand homes in Shaver Lake would be threatened quickly if something was not done immediately. The flames came out of the San Joaquin River Canyon where they had jumped the river to the north the day before, and these flames were sucked up the canyons of Stevenson and Mill Creeks. All that was between a wall of flame fed by hundreds of dead trees per acre and manzanita brush a dozen feet high was Mt. Stevenson. Fortunately for the town of Shaver Lake, this block of land was held privately until sold to Southern California Edison Company around 1920. It was now part of their well-managed forest surrounding the town. Fifty years of rotational logging, broadcast burning, and brush management had produced an open forest environment that made it difficult for fires to crown. This buffered the town from the unmanaged national forest lands on the backside of Mt. Stevenson, which were now ready to be consumed by another flaming head of the Creek Fire.

While the intense, crowning fire moved quickly to threaten the west side of the lake, its first morning's head spread much more slowly on the east side. Entering SCE land about 3:00 AM on the 5th, it burned at a moderate rate of speed to the south and east over the next two days. The private lands it now burned into on the opposite side of the lake from town had been managed extensively over the last forty years by a Forest Service dropout named John Mount.

Chapter 7

The Man Who Saved Shaver Lake

John Mount had graduated from Berkeley with a forestry degree years before it became the center of the counter-culture revolution. He did as most did back then with this degree; he went to work for the U.S. Forest Service. He spent but a few years in their timber program, fighting fire like the rest of his peers when the need arose. Soon his free-thinking approach to forestry, often based on historical and prehistorical evidence, brought him in conflict with the old guard of Forest Service foresters. He got reprimanded for planting sugar pine trees where they weren't currently growing, though he established through research that they had grown there before. (Some of these trees he planted still survive near the top of Tollhouse Grade.) He suggested doing control burning in areas before planting trees and intended to alter current logging practices by planting to benefit wildlife. These ideas were so unacceptably foreign to his supervisors that he was ultimately pushed out of the Forest Service and into private practice. The last job John worked on with the Forest Service was a logging sale that abutted the cabins around Camp Sierra. This was in 1965, the last time the area was logged before the Creek Fire was ignited there.

In business for himself, with a partner, he did blister rust eradication to protect young sugar pine saplings. Shortly thereafter, he was managing several smaller private forest properties for their timber, wildlife, and water. His way of doing things worked, and fire was his friend and an important part of his operation. I went to work for him for the first time in 1972, piling brush on land that was once owned by Herman Peterson, an important figure in local history. I worked summers, holidays, and weekends when he needed me, learning my timber-falling skills on his job the year I graduated high school. His ideas became my ideas early on, and when a short forestry class was offered at my high school, I jumped in.

It was taught by an agricultural education instructor who asked us to submit a study proposal based on an aspect of forestry that interested us. Upon his approval we would then do research and present our findings as a written and oral presentation on our topic. My topic was the effect of controlled fire on wildlife before, during, and after logging. I had firsthand knowledge through hands-on experience from working for John. My proposal was denied. I checked out of the class. The instructor probably worked for the Forest Service at one time.

The results of John's innovative approach to forestry allowed him to secure the management position of over 20,000 acres of SCE-owned forest in the Shaver Lake area. Breaking these lands into management units, he began a rotational treatment schedule that is still followed today. The only forced modification to this plan was during the bark beetle infestation, which had limited effect on SCE lands, and the Creek Fire. These two self-inflicted national forest management blunders, which first turned the Sierra National Forest brown from tree mortality, then black from dead-tree-fueled fires, only marginally affected the well-managed SCE forested property where the Creek Fire was now burning.

The initial fire front stopped at SCE land north of the lake the first morning of the fire and firefighting focus shifted to the west. Meanwhile, the fire continued moving toward the east along this boundary, more quickly on national forest land, less so on SCE's.

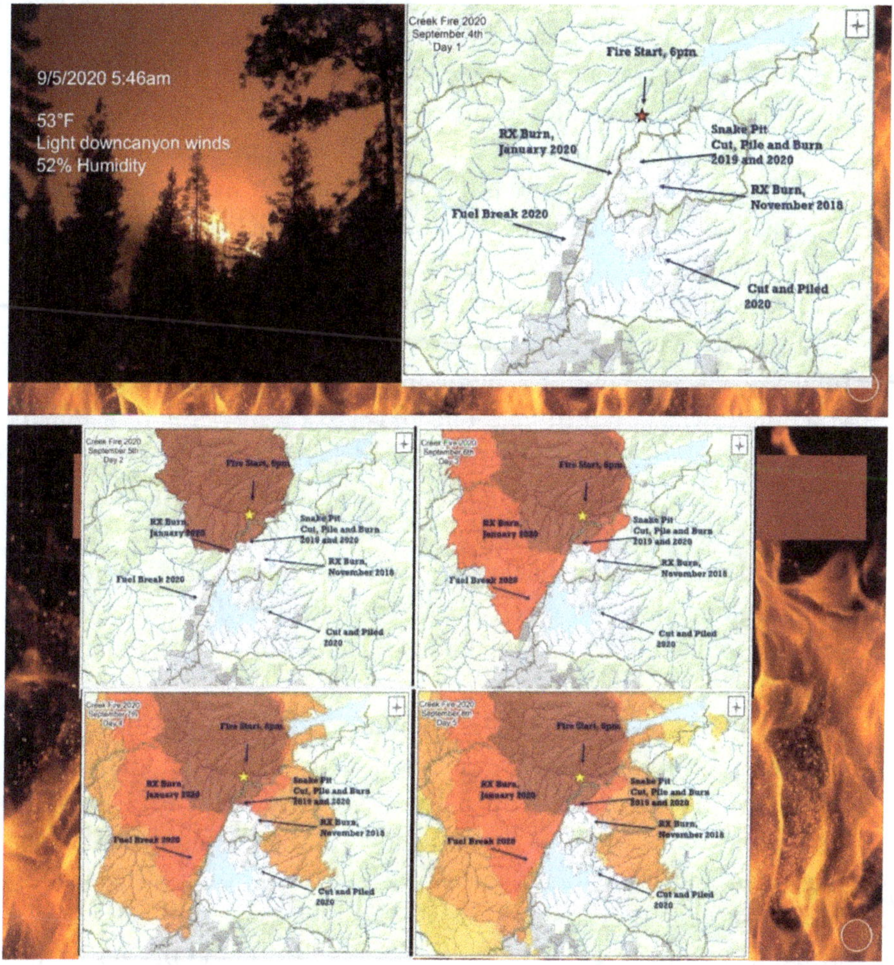

Map of Shaver Lake area with Southern California Edison properties in white and USFS in green. Fire control measures recently put in place by Southern California Edison's forestry department proved crucial in slowing the fire and allowing containment. The 2020 fuel break west of Shaver Lake possibly would have contained the fire as it came up the back side of Mount Stevenson but this chance could not be taken and backfires were set to the east to assure protection of the town of Shaver Lake. Credit: Ryan Stewart

This fire front was insignificant in the big picture that day. It was not putting up several-thousand-foot columns of smoke and nothing to speak of in the way of improvements were threatened. The flames would have to travel miles to the south to skirt the lake and get into Blue Canyon, from which Shaver would then be threatened.

This flanking fire could build little head as it was burning mid-slope with limited fuels above. It finally worked its way over the ridgeline through SCE land as a ground fire where it would have been easily contained if planes, equipment, and manpower were available. Unfortunately, at this time the communities of Big Creek and Huntington Lake were burning, along with the San Joaquin River Canyon. The towns of North Fork and Bass Lake were being threatened along with the outlying area of Cascadel. All assets were focused elsewhere. The fire smoldering through SCE land was getting little attention, but that would change shortly. With its slow movement downhill, a condition that generally slows a fire's advancement, it would take a while to reach Section 16, a rather infamous square mile of isolated Sierra National Forest land surrounded by managed SCE lands.

Section 16 was a forest fire's banquet table set one mile square. It had not been logged in over a hundred years. Parts of it appeared to be an unlogged, old-growth forest which it was not. It was an unmanaged jungle of deep ground fuels and standing dead timber whose animal inhabitants had to access surrounding SCE land to hunt and feed. They could hide very well here but food was scarce. The deer and bear depended on the young brush and berries of the open SCE forest for sustenance. This island of unmanaged fuels had not burned in recorded history and would become a trap for wildlife if they could not escape into the adjoining open forest.

The bark beetle had killed thousands of trees within this block and not a single one of these trees was salvaged before the fire. Once into this unlimited fuel the fire would explode, and that it did. It burned that square mile in a couple hours' time, reaching the southern boundary of the section with so much heat and flame that it crowned into trees well into SCE's fireproof forest, but then it dropped back to the ground. It was contained shortly after this by roads and firelines on SCE property.

If John Mount had not fought the system and found this private forest to showcase his unorthodox forestry practices, Shaver Lake would have burned. It is accepted that managed SCE lands played a huge part in protecting Shaver from the west, but without the same management on the east side of the lake, the fire would have come out of Section 16 and continued nonstop into Blue Canyon. Shaver would have then been targeted by flaming heads from three directions, at nearly the same time. September 6th could have seen Shaver surrounded by flames. Thanks to John Mount it was not.

A Thought

John's management system should be incorporated into plans to protect mountain communities from fire wherever they abut forested lands, whether public or private. Communities of taxpayers should not be at the mercy of negligent forest management. After all, who pays the bills? Millions of dollars spent to protect against fire versus hundreds of millions in firefighting costs, and billions in losses. How does that not pencil out?

John's legacy is well established in people who are continuing his example: Ryan Stewart is now managing the SCE forests, his wife Julianne with Meghan Breniman have taken his practices into their private consulting business, having all worked within John's system and under his auspices. Steve Byrd, Jeff Perini, Riley Young, and John's grandson, Walt Page, are all products of his tireless pursuit of a different approach to forestry and will carry this knowledge into the next generation.

Chapter 8

Timeline (Continued)

Back to September 6th, Shaver Lake

The call was made. Start dozer lines along Highway 168 from the dam toward town and begin backfiring from this line. Backfiring is one of the best methods in controlling an advancing fire. The idea is to have the air being pulled into the advancing fire draw the backfire with it. This would produce a continuous, burned-in fireline along Highway 168. The intent was to do this for four miles; a leap of faith that had no better option. The backfire itself could spot across the line and into the homes and businesses. That is why the pull from the advancing fire is so important. A momentary change of wind and embers would be sent drifting the wrong way, and Shaver could become another casualty of the Creek Fire. Timing was everything and time was running out.

The backfiring operation was successful. First to be saved was the Shaver Lake Point. The lay of the land put it at the top of a chimney where it would be questionable if spot fires into the homes could be managed. Engines were available, having come from all over California and other states, and were placed on patrol to extinguish anything that came across the line.

Of greater concern were the private holdings at the base of Mt. Stevenson. When the land transactions were complete between Southern California Edison Company and the timber property owners early in the last century, an agreement had been made to allow people displaced from their summer homes at the Shaver Mill to be given property nearby, well away from the lake that would soon begin filling. This little community was named Rock Haven and was located just far enough up Mt. Stevenson to have a view of the new lake. The C.B. Shaver home was disassembled in what would become the new lake bottom and reassembled in Rock Haven, where his widow and family would spend summers for several more decades.

Well before the fire front threatened the Shaver Point community, spot fires were erupting in its advance. This spot fire was responded to quickly and the fire extinguished before it could spread into the adjacent lakeside homes. Within two hours the inferno itself approached this end of the lake and raged throughout the night as it burned toward the town of Shaver Lake. Credit: Ryan Stewart

Because access to this group of several dozen homes was marginal at best at any time, concerns were that it could not be saved due to putting firefighters in compromised safety positions. Dozers, both CAL FIRE and private contractors, took their fireline around the historic cabins, and engines were placed throughout to protect against spotting embers. The backfiring from these lines was again successful and Rock Haven survived. Private fuels management instituted by Julianne Stewart and Meghan Breniman within this enclave contributed to the success in saving this historic enclave.

Shaver Lake - Monday, September 7th, 2020

The next few hours were critical for the town of Shaver Lake. Lines to backfire from were put in wherever they could. Some homes would have to be sacrificed. The historic Humphrey home, another timberland holder from the 1880s, was destroyed. The new Crossland home - built atop a cliff - burned, although the old home and honeymoon cabin, also sitting cliff-top, survived. The Erickson home which sat next to a small pond was consumed. These summer homes on Mt. Stevenson meant a lot to me. Our house that I grew up in was just down the road from them. I explored this whole mountain as a child and worked for these cabin owners for years. Many memories burned there.

The backfires and dozer lines did their jobs, along with a recently constructed fuel break atop Mt. Stevenson where the main fire and backfire met. Although Fresno media announced that parts of Shaver were burning - Ken's Market was one establishment singled out by name - nothing had yet burned in the town proper. Those staged at the market, who had just put a night into protecting town, quickly put an end to that rumor. Throughout the daylight hours of the 7th, the fire moved across both the east and west slopes of Mt. Stevenson, burning the isolated home of a close friend of mine as it headed for the Musick Creek drainage.

Safe access to the West Village allowed the numerous engines that had protected Shaver Lake Point to now move into there for structure protection. Soon the fire flanked the mountain and was below them. It was now burning midslope into the chimney my father so feared. Although it did not make a run from The Basin as my dad had foretold, it took hold halfway up the drainage and headed for town. The timing was right for disaster. Afternoon winds were pushed up canyon by valley heat. Was the potential there to build another massive column of suspended embers that would collapse on top of the town of Shaver Lake?

The fire came at the corner of town as expected, generating momentum in old-growth manzanita brush and scattered timber, much of it dead. It was forced to narrow its head the further it traveled, funneled by rock on one side and previously-burned ground on the other. Fortunately, again, Southern California Edison Company lands blocked a portion of its path. This island of private forest was managed specifically as a fire safe area due to its proximity to town with tree spacing much greater and brush nearly non-existent. The fire only had limited opportunity to spread through the thin ground cover under the broken pine, cedar, and fir canopy. These irregularly-connected open areas were debris free and grew scattered, sun-loving plants, which acted as a fire break to the now slow-moving flames. The brush-choked draw to the west, more Sierra National Forest land, did not.

The few available engines were strategically stationed amongst the homes, waiting for the flames to become accessible. The crew members' eyes scanned the sky in search of smoke from spot fires behind them. Well before the fire front reached town, long-range spotting began. It spotted into Ockenden Subdivision and along Bretz Road. It spotted into Musick Heights and West Village. The few engines there were overwhelmed by the number of spot fires. They traveled from one small fire to another, the whole time watching the main front run unimpeded toward town. Volunteers chased spot fires in pickup trucks, putting them out - or at least encircled them with a quick shovel line before chasing another. It was these spot fires that ultimately destroyed several homes in the West Village, but the town was spared. The heavy fuels of the "chimney" were quickly consumed and the fire's momentum was quickly lost. The recipe was not right that day to build a multi-thousand-foot column. Shaver was saved again. One last firebomb awaited detonation below town before it was truly safe - Blue Canyon.

Blue Canyon had not burned significantly since the Bretz Mill Fire of '47, and that was a single run of 3,000 acres that concerned Shaver Lake residents for a day. It nearly went out on its own after leaving the canyon. Two rounds of logging within 75 years had limited the fuels and thus the potential for this fire to spread beyond its

Blue Canyon: October 21, 2015

Blue Canyon tree die off over the course of the summer of 2015. None of these trees were salvaged for their lumber value except along routes of travel as hazard trees. This is what the entire Sierra National Forest looked like at this elevation until the needles fell off the trees. These needleless trees were a major contribution to the energy needed to sustain this fire until it burned into higher elevations and lesser fuel loads. Credit: Ryan Stewart

initial uphill run. Now the Creek Fire, having already burned over 100,000 acres, was heading that direction.

From the Musick Creek drainage below Shaver Lake the fire traveled south after its run toward town, moving through the large-lot subdivision of Dogwood Mountain. Named for the abundance of this beautifully flowered, deciduous, low growing tree, this community should have been immune to the threat of serious fire. Its northern exposure with abundant spring water, the ideal environment for its namesake tree, along with the broken canopy and state fire prevention guidelines required to build there, made it appear to be a relatively fire-safe place to live. This was the case to some extent. One could say the fire randomly picked the structures it burned as it moved through the Dogwood Subdivision toward Highway 168, although this is never the case. Fire does what it does for reasons based on the physics of nature with chance always lurking. One random ember in a fickle wind…and another home feeds the flame.

As soon as the dozers had done all that they could to protect Shaver's west side, they headed blade down toward Bretz Mountain to protect the south side. The dozers worked non-stop for the first four days of the fire near its edge to keep ahead of it. Now they were working in advance of the fire to protect against the consequences of it spotting across Highway168 and moving toward the south. This contingency line would be the last obstacle before the fire reached Blue Canyon. If it jumped this fireline under the current conditions, it would be disastrous. It was up to a handful of men to find a path that would cut off the trail of fuel that could lead the fire into Blue Canyon.

The dozer operators were successful in doing this, and so were the engine crews that held the fire at Highway 168. In several places it jumped this boundary and burned homes, but it was kept from traveling the mile into Blue Canyon. The fire would have one more chance at Shaver.

Jose Basin - September 7th, 2020

The fire had made its way into the upper Stevenson Creek and Mill Creek drainages toward Mt. Stevenson and the town of Shaver Lake after totally consuming the brush and timber of Musick Mountain. It was now advancing on 'The Basin.' Winds were normal for this time of year, light and variable with stronger up-canyon gusts, which was one of the reasons the last run on Shaver did not burn more homes. This same wind pushed the fire at a moderate rate of speed from the head of Redinger Lake toward Jose Basin and the surrounding communities of Big Sandy Rancheria, Meadow Lakes, Mile High, Alder Springs, and Pine Ridge. It kept this pace throughout the day.

The Creek Fire marching across Jose Basin toward Alder Springs the evening of the 7th. This home was one of very few adjacent to Jose Basin that survived the night. Soon after this photo was taken the fire moved into dead trees and thick brush and began building its momentum toward multiple communities. Credit: Ryan Stewart

This front of fire was described as being a mile or more long and unbroken as it swept into 'The Basin' but was not moving quickly or erratically, and was not spotting in advance. The person reporting this thought the fire would be manageable if only some air support was available. This observation changed as quickly as the wind.

From the north came a building breeze toward evening that soon had the fire moving and spotting. The breeze became a wind as the sun set. Its unwavering direction took any erratic behavior out of the mile-wide advancing wall of flame. The head of the fire had now focused its direction, and an ugly black column of smoke and ash was quickly building. Well into Jose Basin and just after dark, the wind-charged fire hit heavy fuels of old-growth manzanita and young Ponderosa pine; the ever-present bomb that locals knew would eventually blow. It is reported that at this time the column began to glow with heat to about three thousand feet above the ground. Lofted, charred fuels were reburning within the column and a collapse was imminent.

The fuel-fed convection produced winds approaching 100 mph into the base of the column from all directions, while the prevailing wind pushed the inferno toward the south. This same wind that moved the fire sent the column of smoke and charred debris over the land in the fire's path. Spot fires hopscotched ahead of the main front and were sucked back into the column's base, feeding the cycle that would increase in strength until fuels became insufficient to support the appetite of the fire.

Fuels began to lessen as the fire burned into the private lands on the southern rim of Jose Basin, and the column came crashing down. In its path was the quaint community of Alder Springs, a town of a few dozen homes and cabins linked by history to the nearby lumber flume. This flume, a 42-mile trough of moving water, sent boards milled in Shaver to the lumber yards of Clovis 125 years earlier.

C.B. Shaver founded a mill in 1892 and arranged the construction of this flume to circumvent the difficulties of hauling lumber 50 miles by freight wagon on questionable dirt roads. The two thousand feet of drop from Pine Ridge to Tollhouse would necessitate tying small trees to these wagon loads of lumber to slow their descent

The Shaver to Clovis flume where it left the rim of Jose Basin and where Pine Ridge School now sits. Lumber from nearby sawmills was loaded into the flume here. Credit: Museum of the Sierra, Shaver Lake

down the infamous Tollhouse Grade. This stretch of road took the lives of several men and many horses and mules before the flume took over the task of moving lumber to the valley floor. A multi-day freight wagon trip might get as much as 5,000 board feet of lumber to Clovis, depending on the dryness of the boards. The flume could easily deliver 30 times this in a twenty-four-hour period.

There were three reasons why Alder Springs was built where it was on a steep hillside over two miles from the flume and the main road at the top of Tollhouse Grade: water, whiskey, and women. The gap where the flume exited the Jose Basin drainage and began its quick descent to Big Sandy Valley was called 'The Saddle.' It is where Pine Ridge School sits now and where C.B. Shaver built a lumber loading access to the flume. This allowed lumber milled in the Pine Ridge area by others and purchased by him to be placed in the flume and shipped to Clovis. Between a lumber-drying yard and other lumbering needs, a small labor force of men was needed. C.B. Shaver, being a devout Methodist, allowed no alcohol or single women on his property. These were the same rules enforced at the mill town of Shaver, where there was a single men's tent-camping area and a married men's area - and alcohol was forbidden. To accommodate these rules at 'The Saddle,' Alder Springs came into existence. Water was abundant, and soon there was a bar, hotel, restaurant, and brothel. A post office soon followed. At Shaver a bar sprang up near where The Point bar is now, just off C.B. Shaver's land, and the 'businesswomen' were found a mile further at another bar near where Rock Haven was later built.

Alder Springs - September 7th, 2020

Alder Springs had been in the sights of the fire all evening, and whether it was the fury of the advancing column, or its direct collapse on top of the community that did it in, Alder Springs was destroyed. Only one small cabin remains of a unique part of local lore.

The fire was nowhere near finished destroying history. As the fire was finishing up its consumption of Alder Springs, it was coming through 'The Saddle.' By 10:00 PM, volunteers and fire crews had done all they could to fire-proof Cressman's Store from the inevitable fire. The hillside above had been cut of brush, and anything combustible that could be moved had been taken far from the historic store and barn area. The fire crews that had been prepping most of the day to save the 115-year-old way-stop were now settled in to steer the fire around the buildings. Confidence was high that this could be done. Combustible ground fuels were a hundred feet away and only a couple of standing, taller green trees remained anywhere near the buildings. All dead trees had been felled and the limbs removed. There was no continuous canopy to crown and nothing for the fire to spot into. Little else could have been done.

For a short time the owner and fire crew watched the glow approach from over the hill, confident the fire would crest the hill in a rush and then slow its approach as it worked its way down toward them. They thought their hoses would keep it at bay until it burned around the well-soaked buildings and went on its way. It was possible the highway might even be a sufficient fire break to stop the fire's advancement into the next canyon, where hundreds of homes were

Cressman's General Store, a way-stop for travelers for over a hundred years, at 10:01 PM, September 7th, 2020. Credit: Ty Gillett

built in very difficult terrain and amongst very heavy fuel loads of brush, oak, and pine.

The fire's approach was heard before it was seen, roaring as if a hurricane, which it nearly was. Next came the flames, soon reaching hundreds of feet into the turbulent sky. The ground fire was not yet visible, its flame blocked from view by distant brush and trees hundreds of yards away. All that could be seen from the Cressman Store were flames reaching high into the night sky. And the heat, building by the second, was unsettling. The fire was several hundred yards from the reach of their hoses, yet the temperatures were climbing rapidly toward the unbearable.

Photo taken six minutes later, 10:07 PM, from Vista Point at the top of the 4-lane. The mass of flame in center of photo is Cressman's General Store burning. The flame to the right is a spot fire between Cressman Road and Peterson Road. By morning the destruction would be finished and over the course of the next several days hundreds of people would realize the fate of their homes; few homes survived in this area. Credit: Ty Gillett

Cressman's General Store, established 1904 - burned September 7th, 2020. Credit: Karen Tillison

With no hesitation, the fire crested the hill and charged the store and its protectors. The flames' height diminished little and the heat became immediately too much for the men to bear. Any uncovered skin would soon burn, but the most frightening aspect of this was the sound of wind and flame, which now had no barrier to block its source. It roared as few things can. Spot fires were already burning to the south and would soon suck back toward them. The buildings could not be saved; those that thought it possible were now trying to save themselves. Air temperatures drove them to their vehicles, and in the last look in their mirrors at a turn less than a hundred yards down the highway, flames could already be seen upon the buildings.

From here the fire went in all unburned directions. This was the collapsing of the column. Thousands of tons of superheated debris rushed back toward Earth under the influence of gravity, displacing air as it approached the ground and produced gale-forced outflow winds. Fire continued south in mile leaps of spotting embers carried by the prevailing wind. All around, outflow winds sent low level fire walls into waiting fuels. By the morning of the 8th, hundreds of homes through Cressman Road, Peterson Road, and Shaver Springs were gone. Our house burned somewhere around midnight, two miles and two hours from Cressman's burned-out shell.

A local logging crew was finishing its day of firefighting while the fire burned across Jose Basin. At about the time Alder Springs was burning, they left the Shaver area for home after a 16-hour day. Highway 168 was closed, but being locals, they knew a back way off the mountain. In two crew cab trucks, one with a 12-foot utility bed, they headed down a dirt road into Blue Canyon. Less than an hour later they had exited Blue Canyon and were on Peterson Road. Flames could be seen well ahead of them and still high on the hill above the road. It appeared they had time to reach Tollhouse Road safely; the thought of reversing course and taking a much longer route home did not enter their exhausted minds.

Two miles later smoke became a problem. Visibility was very difficult on the narrow road. They were now less than a mile from the safety of Tollhouse Road and had not yet encountered flames, just blinding smoke. The idea of turning around was suggested but the narrow road made this difficult in the long-bed truck, so they chose to continue. Around the next bend in the narrow road they had flames on both sides of them and no place to turn around. The fire was moving at them as fast as they were moving at it. They chose to make a run through the fire but quickly encountered small snags that the high wind had brought down. They managed to drive over the first but soon had a larger one crossing their path. The first truck attempted to jump this tree with momentum but only got the front wheels over and was hung up. The fire was now upon them, having spotted below the road and sucked back toward the main fire. They were becoming trapped. The second truck got turned around and the whole crew crammed into it for the escape back the way they had come. Spotting had already ignited both sides of the road they had just passed. Temperatures on the truck's exterior thermometer reached nearly 300 degrees as they split the flames in retreat. They, like those at Mammoth Pool, had survived a column collapse. They would be back at work the next day building firelines.

Less than a mile up Highway 168 from Cressman's, another piece of history was about to be lost.

Armstrong's Hotel and commercial center shortly before it was purchased by the Shaver family and christened Shaver Ranch. The old hotel burned in 1962 and was replaced by a very unique home which burned in the Creek Fire. The Shaver Ranch house has been rebuilt again by the heirs of C.B. Shaver. Credit: Museum of the Sierra, Shaver Lake

A beautiful meadow at the headwaters of Jose Creek was the crossroads of life for the early mountain community. Everything an isolated population needed was found here beginning in the 1880s, up until the paved roads and automobiles made the mountain and valley more accessible. Numerous owners of this property took advantage of the location and ultimately provided food and livery, drink and dance, and a first-rate hotel that later became the Shaver family's mountain retreat. In 1921 the hotel's new name became

Shaver Ranch. Two of my great grandparents' daughters and one of their sons married Bretz siblings after courting at the barn dances and meadow rodeos that were held here. These were important social gatherings for the mountain community and many other relationships resulted from these infrequent opportunities. My grandmother Ruth told me stories of these times, from riding on boards in the flume and eating the blackberries where the flume leaked its acidic water, to horse pack trips into the wilderness where Florence Lake now sits. Our yearly family reunions at Bretz Camp were something to look forward to in my youth, but quite confusing to all of us grandchildren from intertwined marriages decades earlier; we were just cousins of some kind linked by a shared, distant past.

My early mountain relatives standing left to right: Mary (Hauert) Bretz, Ed Bretz, Wilma (Hauert) Bretz, Frank Bretz, Stella Bretz, unknown, Mabel (Hauert) Nelson, Hal Hauert, Lulu (Bretz) Hauert, Frank Hauert. Sitting left to right: Ruth Hauert (my grandmother), Carol Hauert, Bessie Hauert. This photo appears to involve a wedding; don't ask me whose. Credit: Young Family photo

In the winter of 1962 the historic home burned, but quickly a new one took its place. The old hotel's replacement was very unique in design and decor. A lifetime of hunting trophies graced its vast great room and antiques and oddities sat here and there, and everywhere. C.B. Shaver's grandson was the epitome of generosity and many a party was thrown there. I spent my senior year of high school hanging out with a friend that Doug McDonald was boarding in one of the home's many rooms. Hours were spent playing pool or in the sauna and spring fed swimming pool, which happened to harbor huge rainbow trout. The new house could sleep more than the hotel it replaced, and crowds of family and friends were the norm. It was a time I needed as my father had died that fall. The memories remain, but the house, barn, caretaker's cabin and old out-buildings disappeared when the Creek Fire came up Jose Creek, heading for Blue Canyon. Shaver Ranch was temporarily gone, again.

Blue Canyon - September 8th, 2020

Three full days of raging fire and one-eighth of the Sierra National Forest was consumed. By the time winter took care of the last embers, nearly one-third would be burned over by the Creek Fire, much of it catastrophically. There were very few places left to burn around Shaver Lake by September 8th. Four different heads of flame had threatened town and failed. The last threat was, again, Blue Canyon. The fire had already burned in that direction from near Shaver two different times unsuccessfully, cut off by firelines and fuel breaks, both man-made and natural. It was now burning toward Blue Canyon from two new directions, sidehill from Shaver Ranch and from the bottom of the canyon at the end of Peterson Road.

A fire in Blue Canyon was as feared by locals as one in Jose Basin. Over the 70 years since the Bretz Mill Fire threatened Shaver, hundreds of homes had been built atop the rim of this brush-choked drainage. Some of these homes-with-a-view are the most expensive in town, and years of fuel reduction and fire breaks had only marginally limited the fire hazard below them. The morning of the 8th found the fire resting near two edges of the canyon. The general opinion of the evacuated locals watching from a distance was that today would be no different than yesterday, or any day since the fire's start. Because of the destruction wreaked over the last three days, locals' belief was that the fire would eventually wake up with the wind and blow into Blue Canyon. But firefighters could tell something was different. The fire had stopped that morning, after its most destructive run overnight, and was just sitting there smoking. This often happens most mornings on many fires from lower temperatures, higher nighttime humidities, and stagnant winds. How long would these conditions hold? The fire seemed to be waiting to be fought. The opportunity would not be wasted. Planes, engines, dozers, and hand crews rushed to get ahead of it before the fire awoke.

Winds remained absent that day with lower daytime temperatures and humidities. Fire crews were finally fighting a defeatable foe while hoping the change would last long enough to secure the lines. This was accomplished at the mercy of Mother Nature. Another day of heat and wind and the fight would have returned to the edge of town, but again Shaver Lake was spared.

The fire was not done. It would burn for another 123 days and consume another 227,026 acres of forest, but few more structures. It would burn into the far reaches of the wilderness and wait for snow to extinguish its last ember. As it started, it eventually ended as single fleeting spark.

Chapter 9

More Thoughts

I suggested that John Mount's practices could save towns. Cost is the argument always put forth by those that should be doing this on public land but won't. Under the current USFS firefighting system, that could weakly be argued. Yes, it would take millions of dollars to fireproof towns, but few acknowledge the offset of this high price by lessening future firefighting costs and tangible losses, which seem to be accepted by the U.S. Forest Service. On SCE ground, timber sold to mills offsets costs of management. It pays for road maintenance, and it provides employment within the logging industry. The same would be the case on national forest lands, as it was when logging was the means of fire management for a hundred years - not by intent but as a consequence of timber harvesting. In the Shaver Lake area, logging substantially replaced fire in the 1880s by reducing fuel loads and interrupting ground fuels with skid trails and logging roads. By the 1980s, with logged acres becoming fewer and fewer, fire again became the means of reducing forest fuel build-ups, by forest fire on the Sierra National Forest and by prescribed burns on SCE lands. From then to now we have continued to see a drastic reduction in logged acres and the incredible increase in acres burned. Manage the forest properly or watch it burn destructively. The choice is clear.

To make a management system like this work on public lands, the profitability of fighting fires needs to be reevaluated. A fire burning should not be a windfall profit to anyone, but a fire put out could be. Currently, USFS firefighting is a cost-plus business plan. Firefighters get paid considerably more fighting a fire than in station polishing engines and sharpening tools. They need fires to make money, whether they are seasonal employees or full time. How can this be remedied? Pay them the same whether fighting fire or not, and when not fighting fire, they could be doing fire containment projects, fire breaks, and fuels reduction. Instead of being seasonal, pay them to do fire containment projects year-round. Firefighters would now have a full-time job with certain year end wages non-dependent on the fire season.

If not enough of the modern young men and women fresh out of American high schools can do the job of firefighting, bring in the green carded immigrant workers. Make them an army that fights fire and clears brush. Pay them well with incentives to become citizens. Make classwork a part of their job when fires aren't burning. Teach them English and teach them civics so they can pass the test to become a necessary part of our society. Make them an army that treats fire as an enemy that needs to be attacked. Give them direction from generals whose goal is to defeat the fire as quickly as possible without a bureaucracy giving the orders. And give them readily available resources meant for the job. This may end up being a job for the private sector. A bloated bureaucracy always seems to be the goal of any new government undertaking.

Resources meant for the job were not available, or were not utilized, on the Creek Fire. It was a holiday weekend and no USFS law enforcement was available to clear the crime scene - a marijuana grow - so that firefighters could quickly engage the flames. Remedy: Have law enforcement qualifications for at least one member of the fire crew, if not more. This would have allowed the fire to be approached at 7:00 PM instead of 8:30 PM. It also would have allowed the USFS helicopter with bucket to drop on the fire shortly after 7:00 PM. This helicopter never dropped water on the fire. I believe this is because the pilot knew of the possible threat from undesirable people on the ground. One well-placed bucket drop feet above the snag would have eliminated the threat of fire spread and several more would have extinguished the fire.

Two drops of 1,000 gallons from the Sikorsky heavy lift helicopter from 200 feet above the snag were a waste of time, money, and water. In dry air and heat, these long drops can be ineffective. The water evaporates before reaching the ground, or more likely, misses its target. This type of ship picks up water from lakes through a short suction hose, which means it must nearly sit on the water to fill. This limits its usefulness to larger bodies of water from which to fill. A Chinook can fill from any water source that it can submerse its bucket into, since the bucket is attached to the ship by a 200-foot cable. A Chinook would have dropped water from feet above the snag and not only extinguished the flame, but likely knocked the rotten tree to the ground. Sikorskys should not be used on fires in timber; they should only be used on brush and grass where the drop is close to the flame. Remedy: Give Sikorsky helicopters the option of a long line and bucket; it goes wherever the ship goes and can be attached as needed.

Air tankers were in short supply on the Creek Fire partially because of the many fires burning at one time. When fire suppression is nearly entirely dependent on aircraft because of safety protocols limiting ground crews' and equipment's approaches to fires, then these aircraft should never be in short supply. Acquire more aircraft by any means. If there is truly no money for more planes - find it. Downsize the bureaucracy and invest in aircraft. This downsizing of management personnel would streamline the operation and make money more available.

An example of the USFS bureaucracy: The phone call made by the Incident Commander the night of the fire to 'the office' I believed went to the district ranger. Her FOIAed phone records did not show this. Upon investigation of the fire command hierarchy I discovered that the phone call could have gone first to the Assistant Fire Management Officer, who then could have called the District

Fire Management Officer, who then could have called the District Ranger, who then could have called the Forest Fuels Management Officer, who then could have called the Deputy Forest Fire Management Officer (Chief 2), who likely would have been obligated to call the Forest Fire Management Officer (Chief 1), who then would have spoken with the Sierra National Forest Supervisor, Dean Gould.

I do not know the decision-making path that culminated in the 8:32 PM phone call that allowed two hand crews and a dozer to approach the fire. It is easy to see why the decision took over an hour to make with a chain of command like this. That hour wasted allowed the fire to go unextinguished long enough for the wind to blow it into a 380,000-acre conflagration that destroyed a billion dollars in property and made thousands of lives difficult. Is there any question where enough money could be saved to buy another air tanker or helicopter? This is just one national forest. There are 155 in the country.

My last suggestion. Many firefighting helicopters fly all night now. That may not be possible in timbered mountains, but is it possible for them to fly until dark? The privately contracted Sikorsky was on its way to hangar by 7:20 PM, well before dark. A couple of more well-placed drops might have been all it took to put out the piddly-ass fire the Creek Fire was.

Afterthoughts

I have been asked how I got involved in this and why I kept after it for so many fruitless years. They were not fruitless. I took every new discovery as a piece of the pyramid I was building. How long it would take was not a consideration. I told my wife, Dina, numerous times that a new piece of information would force a proper investigation. It would be out of my hands and into someone else's. This still has not happened. I never thought that it would take five years to begin seeing the last pieces fall into place. But also, I never thought that I would be up against corrupt individuals, agencies, and a political machine that runs local politics.

The reason I got involved is much more complicated than just doing what is right. You see in my writing my love for the mountain community and its history. Sixty years in this community formed hundreds of unbreakable bonds - both human and spiritual - with people and places. With the first photo of smoke texted to me the evening of the fire, I became a resident again. Having moved to Montana the previous year, I distanced myself from the daily happenings of the mountain. That little flameless fire took me back emotionally to my roots. A bit of smoke in a bad place and I was young again and wanting to help.

This first text and photo began a stream of information coming my way, and soon it would become a flood. I made a point to respond to everyone and ask them to keep me abreast of the fire. By Sunday the 6th, I knew from multiple sources of the grow evidence near the fire. By the 8th, I knew who likely lost their homes, but did not know who specifically. That included us on Peterson Road. Hours online and on the phone, with my wife monitoring Facebook and other platforms, gave us access to information that those displaced by the fire could not expect to find under their circumstances.

Through an evacuated neighbor of ours on Peterson Road who knew a firefighter dispatched there, we got the photo of their log home that once stood at the end of their driveway. It was gone. In the background of this photo was nothing but devastation. Nothing rose above a couple of feet above the ground except the charred skeletons of oak trees—the oak trees that shaded our front yard from the summer sun. That is how we found out our home had burned.

Through various means, like the video I received from a homeowner that took back roads around roadblocks to view his house, I discovered information. He found his home gone. From his video I was able to identify both standing and burned structures and notify their owners. This one video alone allowed me to tell two families their houses had burned. One of these was my first house, built by myself and friends in my early twenties when my first child was still in diapers. I cherish a picture of him and me sleeping in the antique travel trailer that was home until the roof was on the house. I am not certain if this photo was destroyed in the fire, but it's likely. We still had a lot to move to Montana when the fire hit and are still discovering things we had left that burned.

The hardest to tell of their loss was a very close friend, John Harshman, Jr., whose father was my grandfather's closest friend. Our family ties go back to 1945 when my grandparents bought the store and resort, Ashtonia, in the middle of Shaver, changing the name to Ivie's (their last name). Just down the street was a well-established business opened by John and Velma Harshman in 1927. These were John's parents. The business was called the Trading Post and started as a false-fronted tent, which from the dirt road in front, looked to be a saloon. Through the wooden door was nothing more than a WWI canvas tent. Some of what was sold there came from Herman Peterson's cookhouse. This logger/lumberman had the first train on the mountain, which he used to move timber from near Blue Canyon to his mill. Peterson Road is named for him with part of that road being the original railroad bed.

Trading Post 1926 - Present. Credit: Museum of the Sierra, Shaver Lake

Herman was now into his waning years and was shutting down his logging and milling operation on the north fork of Musick Creek, a trickle of a creek originating in the middle of the town of Shaver Lake. His leftover stock of canned goods was given to the Harshmans to sell. There were many other things the Harshmans made or acquired. Soup, stew, beans, homemade bread, and elderberry wine graced their tables. This basic fare fed much of the mountain community. The main course may or may not have been beef; local venison was often on the menu, as were locally caught fish. Cases of Herman's canned peaches provided many desserts.

From video posted online by a fake journalist who was allowed to roam the mountain using a news pass issued by the Sheriff's Department, I saw that the Harshman home had burned. I could not bring myself to call John and tell him. I called his son to break the news. Their home was like a museum, and his life revolved around history, like mine did. He would be devastated. I still dwell on what was lost there. These following stories come from the hours spent talking to John Harshman, Jr. over a brandy at his and his wife's beautiful Dogwood home.

John Harshman Stories

Tens of hours were spent with a drink in my hand listening to John's stories. I just had one drink per visit, you understand; the glass was never small and was always full at the start of a story. My family's long history on the mountain, some of it shared with his family, made for good conversation. I could ask relevant questions when I felt the need, otherwise, I just listened. When I did ask a question, he was rarely without a reasonable answer. He was a man of his word even while telling a yarn. Occasionally, John would not have an answer and he would say, "Ya know, Jeff, I don't rightly know," and then be back to the story without a second thought. The next time I showed up he would often say, "I gave that some thought," and would fill in the details of the previous story; and then off he'd be on another one, often an offshoot of the last.

The Bear in the Mirror

Dinkey Creek Road was not much of a road until the big mill was built. The dirt path started in the 1880s when a bit of gold was found in Dinkey and Laurel Creeks and in some exposed quartz veins in the Providence Creek drainage. It was a horse-and-wagon affair until a small summer community began to grow around Dinkey Creek, and later, around the sawmill. Then modern means of transportation became the norm. More of its traffic was wide-bunked log trucks hauling timber to the mill than cars and pickups, at least until hunting season.

John Harshman, Jr.

John was a young man in the mid-1950s when an effort was made to improve the road. He was part of a small construction crew that would widen and flatten what was a very rough thoroughfare. They would be there until winter, so a camp was built to house them for the summer and fall. This was assembled on a flat near Summit Creek, where an earlier WPA campsite sat. (The WPA - Works Progress Administration - was a government agency that put men to work during the Great Depression building and rebuilding public infrastructure.) The camp was an efficient base of operations for the men, consisting of several connected canvas tents on a single wooden platform. The cooking, eating, and hand-washing areas were under one roof, with sleeping tents close by. The outhouse from the WPA camp was still functional.

The men's routine would generally start with a shave and wash-up before breakfast, then lunch-making. They would not return until evening. There was a partition between the cooking and eating areas and the washroom, but no door. The only door to the tent was a canvas flap which was generally left open during the day. The washroom amounted to a basin atop a small table with room enough for soap, razors, and a few other toiletries. A mirror hung above and a bucket of creek water sat on the floor below.

One evening the men came home to disaster. The cook tent was nearly entirely on the ground and one wall was split floor-to-ceiling. An investigation began by righting the structure and refastening the supports. With this done the crime was easily solved. White tracks of flour walked about the floor; the flour was from the large container in the kitchen that had supplied it for bread and pancakes. These tracks led to other more edible things, like canned fruit, beans, potatoes, and onions. The remains of these mixed with the flour to form what would have been quite a mess, except this mess had been licked relatively clean. The tracks led from here to the washroom; plain as day like tracks in the snow. The flour-footed bear left regularly-spaced, strolling tracks across the floor to the partition - where he looked into the washroom. What he saw scared the living daylights out of him and he went through the canvas wall in one frantic bound, taking the tent down with him. The terror he saw was nothing more than his own reflection in the mirror.

Afterthoughts, Continued

On the 15th of September the narrative changed about the cause of the fire; I remember that day well because Kamala Harris and Gavin Newsom invaded a friend's burned-out property across from Pine Ridge School. They were there to take advantage of another photo op at another disaster, just to promote their political aspirations. It made national news. The look of empathy on a politician's face is just as likely the look of discomfort at being away from their pampered, coddled life. These two are the poster children for this fake concern. Also, that day lightning was mentioned as a possible cause. The former disgusted me, the latter floored me. Firsthand information was now being ignored in lieu of an impossibility - lightning.

I was now aware of a possible coverup. I began to dig, and found all discoveries not only pointed toward a coverup, but a concerted coverup, and the chance of lightning went from none to contrived. Over the next year, hard evidence was found, and the act of getting an agency to look at what our group had discovered became impossible. The lead USFS investigator told me one Thursday, after following my directions to evidence of a grow, that he had names of possible growers, and by the next Tuesday would not communicate with me. He never picked up the evidence we offered and soon had quit the USFS, scrubbing the internet of his existence. Much the same thing happened with the local sheriff when he assured me that he would pick up the evidence we offered and open an investigation "tomorrow." It was two months before I heard from him again, and that was in an email telling me to get on with my life and let the mountain folks heal. These two interactions, just days apart, were confirmation of a coverup by two separate agencies, in my opinion, attempting to hide their failures, and, also in my opinion, a confirmation of these two separate agencies being in collusion.

Over the next three years information came in sparingly, but what did was explosive. I discovered that numerous pieces of evidence of a grow were seen by firefighters before they were chased out by extreme fire conditions. This evidence was gone when we first accessed the fire four months later. I was able to come up with witnesses that placed the passing of the speeding white vehicles at around 5:30 PM, an hour before the fire was 'officially' reported. I also received information about a list handed to USFS Law Enforcement of questionable people, with their descriptions, coming and going from the area; of shovels and propane tanks on a quad; of voices and small engine noises coming from the area in the middle of the night prior to the fire; and of people seeing buckets, jugs, and drip line near the point of origin of the fire. None of this made the official USFS report, which was signed by the investigator that avoided me for weeks before quitting.

During this time - which was just as telling as the first-person reports - there was a lack of response to questions regularly asked by me of the agencies involved. Hundreds of texts, calls, and emails were made, and nearly none were returned. An occasional response was given when it served them, like identifying the person whose photo had been put on Facebook and labeled the Creek Fire arsonist and grower. The photo was from Lake County the previous year. The sheriff sent me the newspaper report the photo had been taken from. Another time I found a Facebook post revealing the first fire fatality, posted by a local fire battalion chief. I sent this to the sheriff, who had quickly changed the cause of the man's death from fire related to natural, even though the man was trapped in the Sierra backcountry and was awaiting rescue. What he sent me to support that decision was not sufficient to convince me. My own snooping discovered a first-hand account of the man's demise. He was in poor physical condition and had prior heart-related issues. He should not have been at 7,000 feet elevation but had not appeared stressed by his immediate situation. The coroner would have taken all of this into consideration in his findings of natural-caused death by heart attack, but was the thick smoke he was breathing also considered?

Another of John's Stories: Next Year's Firewood

Bert Allen was a character. He died before I was born, but that did not keep me from knowing him. He drank way too much according to everyone but himself. His defense - others did not know how to drink properly. He was found one morning in his '34 Ford by the local bar owner, chest deep in the waters of Shaver Lake. Bert had left at closing time and driven less than a half mile before taking a nap and coasting into the lake. The bar owner always went for a drive after closing and discovered Bert peacefully sleeping. He was cold but fine when awakened and the car would dry out and run for years more and share in other adventures. My favorite story about Bert involved horses, a tall tree, and dynamite.

Bert had a hunting camp above Shaver Lake that he spent a considerable amount of time at, even when it wasn't hunting season. It was not a far walk from the end of the road, a couple of miles at most, but a pack train of horses was necessary to get the supplies in. He and his friends ate very well, and drank even better, and there was always a chance that a deer might need to be packed out. This was a slim chance as the boys rarely left camp.

The camp was located by a large slick rock called Little Baldy. It just happened to be built - I found out during the story - on a small flat where my great uncles' sawmill used to stand, the Huaert-Nelson mill. These were brothers-in-law of the Bretz brothers; the Hauert half was my grandmother's brother who married a Bretz sister, and the Nelson half married my grandmother's sister. One of my cousins of some sort, that I would see yearly at the Bretz family reunion, was born at this mill, supposedly the first birth on the mountain outside of the native Mono Indians. I am still lost as how to label this mixed-up mass of relatives that reunited yearly. We were cousins - cousins of some sort removed somehow. But back to the story.

The camp was well-outfitted with a table and chairs constructed on site, a tent platform that served as the base for the cook tent, a small wood cookstove, and plenty of flat ground to throw out bed-rolls. There was also a large rock fire ring that was obviously very important due to its location; smack dab in the middle of camp. All of the amenities were left at camp after each hunting season, and those which could use some protection were piled under the tent canvas and covered with oiled tarps that did double duty as ground cloths for sleeping. Also stashed in this cache was dynamite, blasting caps, wire, and climbing spurs.

Now, climbing spurs are strap-on-your-leg steel spikes. They allow one to climb trees quite effectively where limbs are not present. They are quite dangerous to use as they regularly slip out of the bark that was meant to hold them. To be safe they are to be used in conjunction with a belt and safety rope that passes around the tree's circumference and is moved up as one ascends and down as one descends. If the spurs fail, the rope attached to the belt - which attaches to you - catches you and keeps you from plummeting, but at the expense of the skin of your nose which has just been driven into the tree. Many a soul has departed Earth due to climbing spurs' unsafe use. Not in Bert's stash was a belt and safety rope.

John Harshman was honored with an invite to the hunting camp, the last trip of the season. He enjoyed the time cooking, gathering wood, and feeding the horses. He even saw a couple of deer while doing this. I asked if he could have shot one, and he replied, "And ruin a perfectly good hunting trip?" The small group of men hunted for the better part of a week without seeing many deer. Poker helped them overlook the lack of game, and when the libations ran out - deer or no deer - the hunt was over. Breaking camp was not much of an ordeal. Trash was burned and gear stowed under the collapsed tent, covered with tarps, and roped to the floor to protect it from the wind. The last thing to do before riding out was secure some wood for next year's fire.

This was accomplished by having the others ride a short distance from camp and tie the horses while Bert climbed a large sugar pine to near its top, without belt and rope. He grabbed the bark with his hands while stabbing it with his spurs, climbing this way until the first limbs were encountered; a distance that would likely send him to his maker if his spurs slipped out. From there it was an easy scamper toward the top of the tree through the large limbs. It was evident he had done this before - the top twenty feet of the tree was limbless. He began working his way down the tree, pulling sticks of dynamite from his pockets and attaching them, with blasting caps, to a half dozen large limbs with twine. He then wired them together for detonation and tossed the coil of wire to the ground, where one of the hunters strung it out to its end. Bert seemed quite comfortable in the limited distance the wire provided from the chaos that was about to erupt. The others, knowing his reputation, took the horses twice that distance and then some. With a hoop and holler, he touched the leads to the battery he had packed in and one great boom filled the air. A couple of pieces of next year's firewood whistled by Bert to land between himself and his campmates. Most fell to the base of the tree. The top of the tree, which was not intended to come down but did, landed squarely on the tent and platform, doing great damage.

Bert just shrugged and said, "I guess I'll have to come in early next year and fix things." To him it was just another successful hunt.

Southern California - January 2025

The Southern California fires left the country reeling. What could not be imagined happened. Towns burned, not just neighborhoods; large portions of towns. People died, and the survivors would never be the same. This opened old wounds to every person that has ever experienced loss by fire. Days of destruction and the aftermath were 24/7 on every screen. The current reality could not be escaped for those directly involved, nor could the reawakened memories be escaped by those who had experienced devastating fires of their own.

Those in the mountain community affected by the Creek Fire must have felt helpless again. I know I did. The Southern California fires did not have to happen. Mistakes of every kind were made; glaring mistakes of insufficient water, rekindling fire, and inadequate warnings. Although different, the Creek Fire mistakes were just as troubling. And in both places the blame could be placed on bureaucracies not designed to do anything more than benefit the bureaucrats. And in both places: dishonesty. Something in those fires triggered others to come forward with information about the Creek Fire. This new information, and a discovery of my own in a photo, put wind in my sails.

From the first days of the fire there were varying rumors of the fire starting at an eradication of marijuana plants by either the Forest Service or the Fresno Sheriff's Department. It was even said that this was done while burning the plants, so no one would have to guard them until they could be flown out by helicopter after the holiday weekend. I could not believe this and always thought the fire was directly related to the growing and harvesting of plants, not a government-initiated eradication.

That spring I received a photo that showed the small fire from a different vantage point and later in the evening than those I had seen before. It was taken just before dark and showed white lights in several places. It also showed the fire burning hot and orange in one small area with smoke moving from it to the northwest. I could place most of the lights; several could be seen as streetlights on Point Road. A few were from Camp Sierra, and above them, from the fire-staging area. To the west was a light I took to be a car on the road to Big Creek. Well below this and feet from the fire, a small white light. Above this, in the area where we would find evidence of the grow, was another small white light. This photo had no timestamp, and it was important that I put a precise time on it. I could not find the provenance of the photo, but kept trying to locate the taker. By chance and not intent, I discovered the time the photo was taken.

On September 4th, 2024, four years to the day of the fire, I sat at the edge of Big Creek heliport with my iPad. Several photos had been taken from here the evening of the fire showing smoke going one way then another. Some were without timestamps. I was here in an attempt to place stars in the sky to compare to those in the unstamped photos. I began my first sweeping video at 6:30 PM; the fire was first responded to at 6:33 PM. I then repeated this process every half hour until 9:00 PM, narrating into the video things of interest, reminders, and thoughts. It was at the 8:00 PM sweep something jumped out at me. A light across canyon had appeared that I had not seen earlier, or never saw move. It was not a parked vehicle but the photocell light over the Canyon Road gate. It was the light in my important photo.

The light came on after 7:30 PM and before 8:00 PM, but how much before? I compared the 8:00 PM video to the photo. It was much darker, nearly black. Sometime near 7:45 PM, I surmised, the photo had been taken. I later established that on that evening the light came on at 7:35 PM and the photo was taken closer to 7:50 PM. I now could say what appears to be at least one, if not two people with headlamps or flashlights were at the fire and the grow at this time. Who were they and what were they doing? I can find no evidence that suggests the Forest Service went back to the fire until after 8:35 PM. The evidence suggests that someone concerned about the discovery of a grow near where a fire was accidentally started, most likely in relation to the grow, was adamant about removing any sign of it.

Photo taken at about 7:50 PM from above Big Creek, September 4th. No one is documented as being in proximity to the fire at this time. USFS engine left fire at about 7:25 PM. USFS Dozer #2 with two hand crews approach fire at about 8:35 PM. Enlargement of fire area shows fire and two light points that appear to be white light, possibly from flashlights/headlamps. Both possible white lights are in areas where evidence of the grow was found by us and other evidence was reported as being seen the evening of the fire by firefighters. The evidence seen by firefighters the night of the fire was gone four months later when I first visited the area. Credit: Museum of the Sierra, Shaver Lake

Evidence of a grow was found by a firefighter at about 8:45 PM and was located about 200 feet from one light in the photo and 50 feet from the other. This evidence directly relating to a grow was in the steep draw leading to a small stream. The most suggestive of several pieces of evidence was the lightly covered drip irrigation that started at the creek. Before the bulldozer arrived, there had been an eradication and cleanup of what the USFS engine captain had found earlier. This was done between 7:30 PM and 8:30 PM. I believe the work was done by sheriff's employees that arrived in the unmarked vehicles which still have not been accounted for. The evidence the firefighter saw at 8:45 PM was gone when I first visited the fire in January of 2021 except for one piece; an empty shot-up fuel can. Someone did not want the marijuana grow to be part of the Creek Fire report, or the Creek Fire history. I will leave you with another story, the last, from John Harshman.

The Trading Post

John, born in 1937, and his brother Loren grew up in the Trading Post, now a two-story building sporting cedar bark siding. Downstairs was business: restaurant, bar, and post office. Upstairs was living area and bedrooms with a bath. The kitchen of the restaurant served both their personal needs as well as that of their customers. Downstairs was a large open room with the bar along one side and dining toward the opposite wall. Between these, against a wall that separated the kitchen from the public area, was a large quartz fireplace. Out front along the dirt road were gas and oil pumps. The outhouses, his and hers, could be seen through the backbar window.

Shaver was never a wild west type of town but frequently gunshots could be heard nearby, usually someone target practicing or shooting a deer or squirrel. But this night there was nearly a shooting in the bar. Going on midnight a man, who had been there drinking for quite some time, retrieved his 30-30 rifle from his truck. The man's intentions had no criminality suggested; he was just showing off his rifle under the influence of too much liquor. Somehow in his showmanship the gun went off, putting a slug through the ceiling above the mantel of the fireplace - and through the floor of John and Loren's bedroom. Panic turned to relief and embarrassment when it was discovered the bullet split the difference between their two beds and continued through the ceiling. I can't remember if John told me if either of them even woke up to all the commotion. The bullet hole can still be seen in the ceiling just above the fireplace.

Meat and cash were in short supply during the depression. Venison could help supply both. John Senior took to the woods in a snowstorm one afternoon in search of a buck deer and soon found a fresh track. Within a quarter of a mile, another hunter's boot tracks merged with the track of the deer. It was snowing hard and John knew that both the hunter and the hunted were just in front of him as the tracks were very clear. He was certain of the identity of the man by the direction he had come, and thus, knew he unlikely had a deer tag. That meant little to John as he knew the meat would not be wasted by this man. If the hunter killed the deer, John would help him drag it out. Shortly a single shot was heard, and soon thereafter John was walking up behind the man, who was bent over gutting the deer. John, knowing the man, spoke his name. Without hesitation, and without casting even a brief glance John's way, the man picked up his gun and knife and ran into the trees, never looking back.

John Senior finished the chore at hand and grabbed an antler and began sledding the buck toward town over the deepening snow. Over the next couple of days he half expected the man to come to claim at least a portion of the deer hanging behind the Trading Post. This never happened. For all anyone knew, the hunter thought the voice he heard was that of the local game warden, and he had escaped the law and would find another deer further from town to poach. John had put meat in the pot without firing a shot. The town ate well for several days, including the game warden, who never questioned where the best stew on the mountain came from.

Some More of My History

My grandparents on my dad's side lived just up the street behind the Trading Post during the war and my Grandpa Young would often sit for hours on the bench out front watching the cars go by and telling stories. He had owned a store and soda bottling works in Dinuba, California, during the roaring twenties and into the depression of the thirties. Connections to my grandmother's family ultimately brought the both of them to the mountains to work for the Southern California Edison Company during WWII. My grandfather was too old for war but took the place of a younger man who had gone off to fight. His work included a job as maintenance man and as a guard on the Shaver Lake Dam, watching for Japanese infiltrators coming by foot or plane. My grandmother did what was available, from cooking at one of the SCE cookhouses to fire lookout duty on Musick Mountain.

In December of '42 my father came of legal age to join-up and headed to the Philippines on the USS Hunter Liggett. My grandparents now had all three sons in the war and Grandpa Young would keep John Harshman, Jr. entertained with stories from across the seas. It was here, at the Trading Post, that my dad courted my mom upon his return from the war.

These small things are important to me. If it were not for history, I would be lost. I think about the reasons for my stubbornness in chasing the truth of the fire, and the one that always rises to the top, above right and wrong, justice and accountability, fraud and deceit, is the history of it all. I cannot imagine my great-great-grandkids still looking at the scars of destruction caused by the Creek Fire, a hundred years from now, and saying lightning caused this. For the sake of history, I hope I have convinced you of the need for truth.

Chapter 10

Politics, Cooperation, and Accountability

I have told the story of the Creek Fire as accurately as possible under the circumstances. What I have written may or may not be the final history of this human and ecological disaster, but the intent was to present it truthfully to the best of my knowledge.

This knowledge was hard to come by. It took five years to get what has been presented, some of the more important information coming in the last year. There are those directly involved with the fire that have knowledge that they have not shared; there are those in positions of management that have discovered this knowledge and have chosen to cover it up.

I cannot leave this project that has consumed such a large amount of my time, energy, and thoughts without addressing the root cause of the Creek Fire Coverup. The fire was not caused by lightning. There was recent human activity at the point of origin. The small fire could have been put out quickly. The fire grew to be the largest in California history from a lack of forest management on government lands. That's the Creek Fire in fifty words or less. The Creek Fire coverup took a book.

Elected politicians had as much to do with this as unelected bureaucrats. My initial concerns were directed toward Sheriff Margaret Mims who showed interest in further information. She had made a public statement that she believed the cause of the fire was not natural. She suggested to one of my contacts that a sheriff's investigation was appropriate if we came up with hard evidence of a coverup. Sheriff Mims quickly handed me off to Undersheriff John Zanoni. There was no suggestion that five years later I would discover she was secreted off the mountain that evening. Her involvement beyond just being there, if any, has not been offered. The need for secrecy has not been addressed either.

My first communication with John Zanoni was about the hard evidence we had discovered (campfire ring, plate, shell casings, and beer can) that came from a hidden campsite discovered by photo analysis of one of the first photos of the fire. He dated the can, as we had done, by the use-by date still visible on the scorched Coors Light can. He did this while on the phone with me, and finding out it had been sold just days before the fire, said he would open an investigation 'tomorrow.' He also told me during that call that he would make arrangements to pick up the evidence and get specific directions to the campsite. I offered to lead a deputy there since it was difficult to both access and locate the small flat the campsite sat on.

This call was made in early January of 2022. I did not hear from him again until March 22, 2022, when he sent me an email telling me he had no interest in the cause of the fire because it had been sufficiently investigated by the U.S. Forest Service. By this time I had sent him numerous emails asking why we still possessed the evidence and he didn't, and why he would not respond to me. By the way, as of May 1, 2026, we still possess the evidence.

Within days of my early January call to Undersheriff Zanoni, I received a call from Deputy Kent Jones asking questions about the fire. This call was 80 minutes of me giving him everything he wanted to know about the fire. I realized quickly by the direction of his questioning that his interrogation was a fishing trip for information. He wanted to know what I didn't know more than what I did. It is now obvious his likely quest was for any knowledge I had of Sheriff Mim's presence at Camp Sierra at the time of its start. I ceased talking by phone with the Fresno Sheriff's Department after this. I realized I needed the written documentation of emails to preserve my communications. The coverup had begun.

At this same time I had made contact by phone with the lead USFS Fire Investigator, Cooper Fouch. I either piqued his interest or he realized that he had a constitutional obligation to look at the evidence we held. I explained by phone where he could find evidence of the marijuana grow and what hard evidence we had gathered and what we left.

Photo taken by fire investigator and texted to me after I directed him to area. This is an impression from a buried 5-gallon bucket. We found two others in this area under bulldozer-disturbed ground. I have video of a group of six of this type of buckets from a previous year's grow fed by drip irrigation line from the creek just to the east of where this bucket impression was found. I believe this evidence was removed between 7:30 PM and 8:35 PM the evening of the fire by individuals associated with either Camp Sierra or the Fresno Sheriff's Department. Credit: USFS by way of text message to me

Two days later I had a texted photo from him of a different bucket impression from the two we had found at the grow site. The text said that he had two possible growers' names from speaking with locals in Big Creek and that he agreed with me that it appeared the USFS dozer had intentionally disturbed the area. He said that the dozer operator was currently working remotely and he would not be able to talk to him for a couple of days. By phone he asked me for directions to the campsite, told me the investigative team never went to that area, and also told me he would visit the campsite then make arrangements to pick up the evidence we held.

That was the last response I had from Investigator Fouch. I believe he talked with USFS Dozer Operator, Jeff Cornell, and discovered the truth of the coverup. For about a three-week period I attempted to communicate with Investigator Fouch. I sent numerous emails, a couple of texts, and several phone messages. None were returned. He soon quit the Forest Service and his internet history was erased. It took me 18 months to find a hint of him on the web - a mention of his name at a fire meeting in Nebraska. Just recently I discovered he is working in law enforcement again, this time with the National Park Service. I sent him a recent email reminding him of his constitutional obligation to do his job correctly and ask if he was interested in sharing his knowledge of the Creek Fire origin. I have received no response.

More of those that have accountability within the U.S. Forest Service: High Sierra District Ranger, Kim Sorini, who has directed questions from me about the fire to her attorney. Sierra National Forest Supervisor, Dean Gould, whose only email response to the score or more I have sent him, was to ask me to get on with my life and let the community and forest heal - neither of which will ever fully happen. Joseph Sadowski, USFS Region 5 LEO. I went here with an offer of evidence after no cooperation at the Sierra National Forest level. My email was blocked. He said the blocking of the email could not have happened. He was sent the screenshot of it. He then accepted my email but never acted upon my request to view the evidence, only telling me he was too busy. USFS Region 5 Agent in Charge, Don Hoang. He has not responded to my emails and has denied FOIA requests. 'No response' seems to be a taught strategy in doing one's job in government.

Besides USFS management's lack of cooperation, there are others in the Forest Service that share some accountability. The aforementioned dozer operator, Jeff Cornell, that likely did evidence coverup at the request of a boss. He also made false statements about why he could not approach the fire sooner. The trainee fire boss the first evening of the fire, Battalion Chief Victor Lefor, who received information from others and turned that information into a firefighting plan that failed. Engine 42 Captain, Robert McDonald, trainee boss that directed his engine to drive down the 'inaccessible' four-wheel drive road to the fire, thus endangering the crew. He then gave false statements that were entered into the official fire report about this access to the fire and the evidence of a grow he saw there. Patrolman 34, Brian Kilcrease, who stated to residents that the fire was not being fought because of a marijuana grow, then told me by phone there was no grow and that I should not insinuate that firefighters were lying about its existence.

Moving on to other Fresno County entities that were not only guilty of lack of cooperation, but of total avoidance of my concerns about the untruthful and incorrect narrative of the fire. The Fresno County Board of Supervisors in their entirety, but most notably Nathan Magsig, District 5 Supervisor, who firmly believed the fire was not started by lightning but would not assist his constituents in discovering the true cause. Fresno County Grand Jury Forepersons, four of them, who have passed their individual responsibility on for several years now. And the most egregious of all, District Attorney Lisa Smittcamp, whose lack of duty, responsibility, and constitutional obligations are only eclipsed by her total unprofessionalism. I asked her to investigate, without response, and then sent her by email evidence we had found with details, names, dates, and leads. I gave her information beyond what it should have taken for an inquiry. I gave her a canned investigation with photos and offered radio recordings, FOIA materials, and my personal assistance on the ground. She would not reply. I reminded her regularly of her constitutional requirements to investigate a crime, without results. District Attorney Smittcamp ultimately requested a friend of mine to tell me not to email her anymore. How unprofessional can it get? The top law enforcement individual in the county cannot speak for herself, let alone do her job. Fire her, and John Zanoni, at the next election. And Nathan Magsig wants to be a state senator?

Fire	From: **September 4, 2020** To: **May 25, 2021**
Case Status **Closed—Solved**	Report Type **Final**
Case Name **Creek Fire, SNF**	Investigated By **Special Agent** b6 & b7C

SYNOPSIS

On September 4, 2020, at approximately 1818 hours, Fresno County Sheriff's Dispatch received a 911 call indicating there was a fire in the area of Big Creek, CA near Camp Sierra. Additionally, the fire was called into the Sierra National Forest Dispatch Center at approximately 1833 hours by a Fire Prevention Technician stationed in Big Creek.

The Creek Fire originated just south Big Creek drainage, west of Camp Sierra and north of Huntington Lake Road, on the Sierra National Forest, within the Eastern District of California, Fresno County. The Creek Fire burned 379,895 acres, destroyed 853 structures, and damaged an additional 64 structures.

To date, investigators have exhausted all available leads and resources in an attempt to identify subject/s that were in the vicinity of the Creek Fire origin prior to and shortly after the fire was reported. At this time, no subject/s have been identified. Based on the above facts, along with the fact that no physical evidence was located to support any other fire cause categories as the probable cause of the Creek Fire; investigators have determined that lightning is the probable cause **(see exhibit 30, Creek Fire Origin and Cause Report)**.

Future leads or information obtained related to the cause of the Creek Fire may result in a change of this conclusion.

Case considered closed, pending further leads.

Distribution

Unit		Copies
LEI, WO		
LEI, Region	5	1
LEI, Region		
US Attorney		

Prepared By: b6 & b7C — Digitally signed by b6 & b7C Date: 2021.07.15 11:43:03 -07'00' — Date

Reviewed By: b6 & b7C — 07/15/2021 — Date — 07/15/2021

The first page of the Creek Fire Report of Investigation. The reason that the 6:18 PM report of the fire to the Fresno Sheriff's 911 dispatcher was not relayed to the USFS is never addressed. The other obligations of responsibility stated here were never pursued by the USFS. I believe an independent reinvestigation of the cause of the Creek Fire is called for due to this negligence. Credit: USFS via internet

The local media that helped: I will start with the only journalist in the whole valley that gave me the time of day: Marek Warszawski. He initially bought into the misinformation but soon discovered the unlikelihood of lightning being the cause. He asked locals of the possibility of marijuana grows in the area and was told of two the sheriff's department had eradicated in the past, both very near the Creek Fire's point of origin. He questioned the sheriff about these marijuana grows and asked for records of them and other grows in the area and was denied. At this time the sheriff was Margaret Mims. She would soon resign and appoint Undersheriff John Zanoni to her position. Marek worked with me toward the truth, but was limited by his employer, McClatchy Broadcasting. They did not want him doing real journalism and would only let him 'tell Jeff's story,' which he did. My thanks to Marek for his interest in and attempt to expose the truth. The biggest help, and the one I could not have done without, was the Mountain Press. A small, local paper that printed my letters, thus allowing me to communicate with the mountain community from distant Montana. Immeasurable thanks to the local community who gave their unwavering support and assistance and who are the reason I have gotten this far.

The media that didn't help: All the rest of the valley print and broadcast news. I communicated with all of the big names in San Joaquin Valley journalism without so much as a reply, until I shamed Dominic McAndrews (Channel 30 Fresno) into contacting me. I did this by telling him that the amount of effort the station put into disproving an internet hoax would have gotten to the truth of the Creek Fire. With this coaxing, he agreed to accept a copy of the official fire report (ROI). I sent him this with my notes addressing the misinformation, false statements, and omissions that were throughout the document. I spent six hours of writing doing this. He spent sixty seconds in telling me he wouldn't look into it.

I sent an email to Sheldon Gajarian at KMPH 26, Fresno, requesting his assistance in getting word out to the county residents of the intent to petition the Fresno County Grand Jury to do an inquiry. This inquiry was about the uncertain participation of the

Fresno Sheriff's Department the first evening of the Creek Fire. His response was not only no, he added that KMPH would not be involved in anything I was a part of. I believe not only Sheldon Gajarian and Dominic McAndrews were visited, but others had been talked to about working with me by one or more of the agencies involved, most likely the Fresno Sheriff's Department. I guess that's what to expect from journalism today. A side note from when I was in the teaching program at Fresno State: In a class on teacher's ethics, we were given a handout on cheating in college. Of the multiple disciplines, engineering had the lowest frequency of cheating at 42%, teaching was near 50%, and journalism was the highest near 66%. Go figure.

May my persistence pay off in the truth.

Thanks for your time and I hope you found my book enlightening, if not entertaining.

—Jeff Young

About the Author

The fourth of six generations that have called the Fresno County mountains home, Jeff left at the age of 60, with his wife, Dina, to Montana for a less cluttered, traditional environment. One year later the Creek Fire burned the forest he had explored since childhood, and worked in as a timberfaller his entire career. This fire also destroyed their home they had yet to sell, as well as dozens of friend's homes. Long held relationships with people trusted for decades produced information contrary to what was presented by fire, law enforcement, and media, thus causing an overwhelming curiosity and desire for the truth as it related to the start and extinguishment of the Creek Fire. Out of respect for the mountain community that I will always be a part of, I committed to discovering this truth.

www.ingramcontent.com/pod-product-compliance
Lightning Source LLC
LaVergne TN
LVHW022229200926
848481LV00006B/2193
* 9 7 9 8 9 9 6 8 7 3 6 0 9 *